Advances in Intelligent Systems and Computing

Volume 537

Series editor

Janusz Kacprzyk, Polish Academy of Sciences, Warsaw, Poland
e-mail: kacprzyk@ibspan.waw.pl

About this Series

The series "Advances in Intelligent Systems and Computing" contains publications on theory, applications, and design methods of Intelligent Systems and Intelligent Computing. Virtually all disciplines such as engineering, natural sciences, computer and information science, ICT, economics, business, e-commerce, environment, healthcare, life science are covered. The list of topics spans all the areas of modern intelligent systems and computing.

The publications within "Advances in Intelligent Systems and Computing" are primarily textbooks and proceedings of important conferences, symposia and congresses. They cover significant recent developments in the field, both of a foundational and applicable character. An important characteristic feature of the series is the short publication time and world-wide distribution. This permits a rapid and broad dissemination of research results.

Advisory Board

Chairman

Nikhil R. Pal, Indian Statistical Institute, Kolkata, India
e-mail: nikhil@isical.ac.in

Members

Rafael Bello, Universidad Central "Marta Abreu" de Las Villas, Santa Clara, Cuba
e-mail: rbellop@uclv.edu.cu

Emilio S. Corchado, University of Salamanca, Salamanca, Spain
e-mail: escorchado@usal.es

Hani Hagras, University of Essex, Colchester, UK
e-mail: hani@essex.ac.uk

László T. Kóczy, Széchenyi István University, Győr, Hungary
e-mail: koczy@sze.hu

Vladik Kreinovich, University of Texas at El Paso, El Paso, USA
e-mail: vladik@utep.edu

Chin-Teng Lin, National Chiao Tung University, Hsinchu, Taiwan
e-mail: ctlin@mail.nctu.edu.tw

Jie Lu, University of Technology, Sydney, Australia
e-mail: Jie.Lu@uts.edu.au

Patricia Melin, Tijuana Institute of Technology, Tijuana, Mexico
e-mail: epmelin@hafsamx.org

Nadia Nedjah, State University of Rio de Janeiro, Rio de Janeiro, Brazil
e-mail: nadia@eng.uerj.br

Ngoc Thanh Nguyen, Wroclaw University of Technology, Wroclaw, Poland
e-mail: Ngoc-Thanh.Nguyen@pwr.edu.pl

Jun Wang, The Chinese University of Hong Kong, Shatin, Hong Kong
e-mail: jwang@mae.cuhk.edu.hk

More information about this series at http://www.springer.com/series/11156

Jezreel Mejia · Mirna Muñoz
Álvaro Rocha · Tomas San Feliu
Adriana Peña
Editors

Trends and Applications in Software Engineering

Proceedings of CIMPS 2016

 Springer

Editors
Jezreel Mejia
Unidad Zacatecas
Centro de Investigación en
  Matemáticas A.C.
Zacatecas
Mexico

Mirna Muñoz
Unidad Zacatecas
Centro de Investigación en
  Matemáticas A.C.
Zacatecas
Mexico

Álvaro Rocha
Departamento de Engenharia Informática
Universidade de Coimbra
Coimbra
Portugal

Tomas San Feliu
Lenguajes y Sistemas Informáticos e
  Ingeniería de Software
Universidad Politecnica de Madrid
Boadilla del Monte, Madrid
Spain

Adriana Peña
Departamento de Ciencias Computacionales
Universidad de Guadalajara
Guadalajara
Mexico

ISSN 2194-5357   ISSN 2194-5365   (electronic)
Advances in Intelligent Systems and Computing
ISBN 978-3-319-48522-5   ISBN 978-3-319-48523-2   (eBook)
DOI 10.1007/978-3-319-48523-2

Library of Congress Control Number: 2016954910

This Springer imprint is published by Springer Nature
The registered company is Springer International Publishing AG
The registered company address is: Gewerbestrasse 11, 6330 Cham, Switzerland

Preface

This book contains a selection of papers accepted for presentation and discussion at The 2016 International Conference on Software Process Improvement (CIMPS'16). This Conference had the support of the CIMAT A.C. (Mathematics Research Center/Centro de Investigación en Matemáticas), CANIETI (Cámara Nacional de la Industria Electrónica, de Telecomounicaiones y Tecnologías de la Información), ITA (Instituto Tecnologico de Aguascalientes, México), AISTI (Iberian Association for Information Sistems and Technologies/Associação Ibérica de Sistemas e Tecnologas de Informação), ReCIBE (Revista electrónica de Computación, Informática, Biomédica y Electrónica). It took place at Instituto Tecnológico de Aguascalientes, Aguascalientes, México, from 12th to 14th October 2016.

The International Conference on Software Process Improvement (CIMPS) is a global forum for researchers and practitioners that present and discuss the most recent innovations, trends, results, experiences and concerns in the several perspectives of Software Engineering with clear relationship but not limited to software processes, Security in Information and Communication Technology and Big Data Field. One of its main aims is to strengthen the drive towards a holistic symbiosis among academy, society, industry, government and business community promoting the creation of networks by disseminating the results of recent research in order to aligning their needs. CIMPS'16 built on the successes of CIMPS'12, CIMPS'13, CIMPS'14, which took place on Zacatecas, Zac, and CIMPS'15 which took place on Mazatlán, Sinaloa, México.

The Program Committee of CIMPS'16 was composed of a multidisciplinary group of experts and those who are intimately concerned with Software Engineering and Information Systems and Technologies. They have had the responsibility for evaluating, in a 'blind review' process, the papers received for each of the main themes proposed for the Conference: Organizational Models, Standards and Methodologies, Knowledge Management, Software Systems, Applications and Tools, Information and Communication Technologies and Processes in non-software domains (Mining, automotive, aerospace, business, health care, manufacturing, etc.) with a demonstrated relationship to software process challenges.

CIMPS'16 received contributions from several countries around the world. The papers accepted for presentation and discussion at the Conference are published by Springer (this book) and, extended versions of best selected papers will be published in relevant journals, including SCI/SSCI and Scopus indexed journals.

We acknowledge all those who contributed to the staging of CIMPS'16 (authors, committees and sponsors); their involvement and support is very much appreciated.

Aguascalientes, México   Jezreel Mejia
October 2016   Mirna Muñoz
                                                                    Álvaro Rocha
                                                                Tomas San Feliu
                                                                  Adriana Peña

Organization

Conference

General Chairs

Jezreel Mejia, Mathematics Research Center, Research Unit Zacatecas, MX
Mirna Muñoz, Mathematics Research Center, Research Unit Zacatecas, MX

The general chairs and co-chair are researchers in Computer Science at the Research Center in Mathematics, Zacatecas, México. Their research field is Software Engineering, which focuses on process improvement, multi-model environment, project management, acquisition and outsourcing process, solicitation and supplier agreement development, agile methodologies, metrics, validation and verification and information technology security. They have published several technical papers on acquisition process improvement, project management, TSPi, CMMI, multi-model environment. They have been members of the team that has translated CMMI-DEV v1.2 and v1.3 to Spanish.

General Support

CIMPS General Support represents centers, organizations or networks. These members collaborate with different European, Latin America and North America Organizations. The following people have been members of the CIMPS conference since its foundation for the last 4 years.

Cuauhtémoc Lemus Olalde, Head of Cimat Unit Zacatecas, MX
Angel Jordan, International Honorary, Software Engineering Institute, US
Laura A. Ruelas Gutierrez, Government Zacatecas, MX
Gonzalo Cuevas Agustin, Politechnical University of Madrid, SP
Jose A. Calvo-Manzano Villalón, Politechnical University of Madrid, SP
Tomas San Feliu Gilabert, Politechnical University of Madrid, SP
Alvaro Rocha, Universidade de Coimbra, PT

Local Committee

CIMPS established a local committee of selected experts from the Cámara Nacional de la Industria Electrónica de Telecomunicaciones y Tecnologías de la Información (CANIETI) and of Instituto Tecnológico, Aguascalientes (ITA), Aguascalientes. The list below comprises the Local Committee members.

CANIETI:

Montserrat Villalobos, Local Chair, MX
José Luis Macias, Public Relations, MX
Gabriel Suaréz, Public Relations, MX
Alejandro López, Public Relations, MX
José Luis Rodriguez, Public Relations, MX.
Guadalupe González, Logistics, MX
Alejandro González Dávila, Public Relations, MX

ITA:

Edinguer Vázquez Ayala, Logistics, MX
Pedro Pablo Martínez Palacios, Conferences, MX
Luis Antonio Cruz Macías, Logistics, MX
Joel Azpeitia Luévano, Logística
Maricela Azuara Dueñas, Conferences, MX
Alejandro Sánchez Barroso, Conferences, MX
Iraam Antonio López Salas, Logistics, MX
Martha Catalina de Lira Ortega, Logistics, MX

Scientific Program Committee

CIMPS established an international committee of selected well-known experts in Software Engineering who are willing to be mentioned in the program and to review a set of papers each year. The list below comprises the Scientific Program Committee members.

Adriana Peña Perez-Negrón, University of Guadalajara CUCEI, MX
Alejandro Rodríguez Gonzalez, Politechnical University of Madrid, SP
Alejandra Garcia Hernández, Autonomous University of Zacatecas, MX
Alma Maria Gómez Rodriguéz, University of Vigo, SP
Alvaro Rocha, Universidade de Coimbra, PT
Antoni Lluis Mesquida Calafat, University of Islas Baleares, SP
Antonia Mas Pichaco, University of Islas Baleares, SP
Antonio de Amescua Seco, University Carlos III of Madrid, SP
Benjamin Ojeda Magaña, University of Guadalajara, MX
Carla Pacheco, Technological University of Mixteca, Oaxaca, MX
Carlos Lara Álvarez, CIMAT Unit Zacatecas, MX
Diego Martín de Andrés, University Carlos III of Madrid, SP
Edrisi Muñoz Mata, CIMAT Unit Zacatecas, MX
Edwin León Cardenal, CIMAT Unit Zacatecas, MX

Elisabet Cápon, Swiss Federal Institute of Technology, Zürich (ETHZ), CH
Eleazar Aguirre Anaya, National Politechnical Institute, MX
Fernando Moreira, University of Portucalense, PT
Gabriel A. Garcia Mireles, University of Sonora, MX
Giner Alor Hernandez, Technological University of Orizaba, MX
Gloria P. Gasca Hurtado, University of Medellin, CO
Graciela Lara Lopez, University of Guadalajara CUCEI, MX
Gonzalo Luzardo, Higher Polytechnic School of Litoral, EC
Gustavo Illescas, National University of Central Buenos Aires Province, AR
Hector Duran Limón, University of Guadalajara, MX
Hugo Arnoldo Mitre, CIMAT Unit Zacatecas, SP
Ivan A. Garcia Pacheco, Technological University of Mixteca, Oaxaca, MX
Javier García Guzman, University Carlos III of Madrid, SP
Jezreel Mejia Miranda, CIMAT Unit Zacatecas, MX
Jörg Thomaschewski, University of Vigo -Hochschule Emden-Leer, SP
Jose A. Mora Soto, CIMAT Unit Zacatecas, MX
Jose Antonio Cerrada Somolinos, National Distance Education University, SP
Jose Baltasar García Perez-Schofield, University of Vigo, SP
Josue N. Garcia Matias, Nova Universitas, MX
Juan Manuel Toloza, National University of Central Buenos Aires Province, AR
Lohana Lema Moreta, University of the Holy Spirit, EC
Luis Casillas, University of Guadalajara CUCEI, MX
Luis J. Dominguez Pérez, CIMAT Unit Zacatecas, MX
Magdalena Arcilla Cobián, National Distance Education University, SP
Manuel Pérez Cota, University of Vigo, SP
María Y. Hernández Pérez, Electrical Research Institute, MX
Marion Lepmets, Regulated Software Research Centre, Dundalk Institute of Technology, Ireland
Miguel Hidalgo Reyes, National Center for Research and Technological Development, CENIDET, MX
Mirna Muñoz Mata, CIMAT Unit Zacatecas, MX
Omar S. Gómez, Higher Polytechnic School of Chimborazo, EC
Paul Clarke Dundalk Institute of Technology, IE
Perla Velasco-Elizondo, Autonomous University of Zacatecas, MX
Ramiro Goncalves, University Tras-os Montes, PT
Raúl Aguilar Vera, Autonomous University of Yucatán, MX
Ricardo Colomo Palacios, Østfold University College, NO
Rory O'Connor, Dublin City University, IE
Santiago Matalonga, University ORT, UY
Sodel Vázquez Reyes, Autonomous University of Zacatecas, MX
Sulema Torres Ramos, University of Guadalajara CUCEI, MX
Tomas San Feliu Gilabert, Politechnical University of Madrid, SP
Ulises Juárez Martínez, Technological University of Orizaba, MX
Ulrik Brandes, University of Konstanz, GL
Valentine Casey, Dundalk Institute of Technology, IE

Vianca Vega, Catholic University of North Chile, CL
Victor Flores, Catholic University of the North, CL
Victor Saquicela, University of Cuenca, EC
Yadira Quiñonez, Autonomous University of Sinaloa, MX
Yilmaz Murat, Çankaya University, TR

Contents

Part I
Organizational Models, Standards and Methodologies

Implementing the Ki Wo Tsukau® model to strengthen the commitment of small-sized software enterprises in software process improvement initiatives

Garcia, I.[1], Pacheco, C.[1], Calvo-Manzano, J. A.[2], and Hernández-Moreno, H.[1]

[1] Division de Estudios de Posgrado,Universidad Tecnologica de la Mixteca,
Carretera a Acatlima, 69000 Oaxaca, Mexico
{ivan, leninca, hmoreno}@mixteco.utm.mx
[2] Escuela Tecnica Superior de Ingenieros Informaticos, Universidad Politecnica de Madrid,
Boadilla del Monte, 28660 Madrid, España
joseantonio.calvomanzano@upm.es

Abstract. In spite that the lack of commitment and motivation has been directly linked to a high degree of failures in software process improvement initiatives, these two aspects have not received yet enough attention by researchers. In this regard, small enterprises are more susceptible to fail when committing an effort for improving because the people involved do not know what to do or how to act with this change. Moreover, the Ki Wo Tsukau® model is promoting the use of proper levels of energy, commitment, and motivation in different contexts. Therefore, this paper introduces an alternative approach to strengthen the commitment of a Mexican small-sized software enterprise, by implementing some basic principles of the Ki Wo Tsukau® model in a software process improvement initiative.

Keywords: Software process improvement, Ki Wo Tsukau® model, small-sized software enterprises.

1 Introduction

Nowadays, quality is one of the main objectives for organizations that provide similar software products, because it will influence the customer's decision and, as consequence, it will provide more economical benefits. Taking into account this situation, organizations are trying to develop better quality products because their survival depends on the quality of products and offered services [1]. In this regard, the rapid change and global competitiveness have originated that worldwide software organizations worry about being more efficient and deliver quality products and services. Nevertheless, this is not an easy task for organizations, in particular for the smaller ones. In spite that the small-sized software enterprises —companies with less than 50 employees that have been independently financed and organized— make a significant contribution to the economy of any country in terms of employment, innovation and growth, they face to many problems when attempting to increase the quality of their products. In this context, it is important to mention that an increasing portion of worldwide ˙software production is done by small-sized enterprises.

© Springer International Publishing AG 2017
J. Mejia et al. (eds.), *Trends and Applications in Software Engineering*, Advances in Intelligent Systems and Computing 537, DOI 10.1007/978-3-319-48523-2_1

Moreover, it is well known that the software production is characterized, among other characteristics, by mainly relying on the skills of the individuals who build the software. Therefore, the standardization of software development processes, by the maturation of organizational processes, has become imperative to achieve significant levels of competitiveness and innovation [2]. According to [3], processes innovation is the implementation of a new method, or one significantly improved, of production or distribution of products and services, including major changes in techniques, teams, and/or software methods. With this aim in mind, it is important to define two key concepts for a successful innovation: the *absorptive capacity* that, according to [4], involves the internal capacity of enterprises for developing knowledge by accumulating experience, and the *learning* that suggests a change in the enterprise's model and, as consequence, the maintenance or improvement of its performance [5]. Furthermore, the learning capacity of software enterprises should be reflected on a change in organizational practices, in order to support or improve the obtained results.

However, process innovation is not easy within the context of the small-sized software enterprises. Research by Psomas, Fotopoulos, and Kafetzopoulos [6], for example, affirms that among the critical factors to promote the process innovation in these enterprises are the following: lack of staff's motivation and commitment, lack of individual skills of employees, and the lack of guidelines to simplify the complexity about how to innovate. Moreover, Ñaupac, Arisaca, and Davila [7] have identified some other factors related to the lack of experience and poor technical knowledge, and lack of commitment and credibility on the process innovation possibility. Similarly, research by Muñoz, Gasca, and Valtierra [8] identified some constraints such as lack of commitment and resources (e.g., financial, human, facilities), lack of understanding the relevance of process on product's quality, and lack of a defined process that substitutes the artisanal labor. In this regard, the lack of commitment is a common factor that has been considerably explored within the context of process innovation in small-sized software enterprises [9-11].

Thus, this paper aims to present the implementation of the Ki Wo Tsukau® model in order to achieve and strengthen the commitment of management and operative levels within the context of the activities for Software Process Improvement (SPI) initiatives in small-sized software enterprises. The rest of this paper is organized as follows: Section 2 provides a detailed definition of the Ki Wo Tsukau® model. In Section 3 our guidelines and advice are presented in order to provide an easy implementation for SPI. Furthermore, in order to illustrate the practicability of our approach the results of a preliminary validation are also provided in Section 4. Finally, Section 5 draws the main conclusions and findings of this study.

2 Ki Wo Tsukau® model

This model can be seen as a continuous improvement assurance model within the context of quality management systems under ISO 9001:2008 [12]. Moreover, the Japanese "Ki Wo Tsukau" term can be understood as "Ki" (energy) and "Wo Tsukau" (to use), and it can be interpreted as "to be worried of, to pay attention to another's needs, to attend to, to fuss about, to take into consideration, or to correctly use the energy". This model was created in Yakult, Puebla (Mexico) by Alejandro Kasuga and Dr. Aldo Trujillo and since 2012 it has been promoted as an international standard of the ISO 9000 series by the ISO Technical Committee 176 – Quality management and Quality Assurance in China, Russia, and Holland.

The Ki Wo Tsukau® model is based on the application of the Deming's PDCA cycle [13], in order to seek the improvement of all processes through proactivity and performance and by focusing on the satisfaction of internal and external customers. Therefore, the model establishes indicators that are not based in preventing unconformities, but in the pursuit of the continuous improvement of processes. That is, unlike the sophisticated and complex tools for continuous improvement that traditionally could be used within the organizations and that can only be applied by a minority (i.e., due to the budget constraints), Ki Wo Tsukau® is based on the involvement of all employees not only to perform activities to fulfill their basic functions, but to promote proactivity through process improvement seeking the customer' satisfaction. In this regard, Fig. 1 shows that the Ki Wo Tsukau® model considers three variables: K1 (representing basic functions performed by an employee), K2 (which is the level of proactivity), and K3 (the impact or benefit in K1).

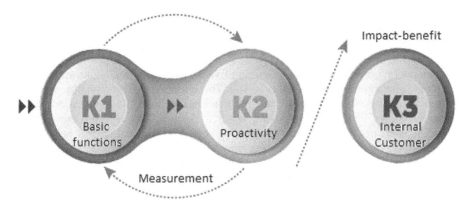

Fig. 1. Variables of the Ki Wo Tsukau® model [14]

In spite that the Ki Wo Tsukau® model has not been specifically created for the context of software industry, according to their creators these variables can be applied in diverse contexts (e.g., industrial environments, government, education) as follows:

- K1: This variable represents all the activities performed by the organization's staff, it means those activities for which they were hired and that may, or not, be documented in a manual, procedure, work instructions, etc.
- K2: This variable is related to the talent and experience of each employee to perform innovative solutions in order to improve the processes and, as a consequence, to obtain quality products. In this regard, this variable assumes that if the knowledge acquired by formation or training has a positive impact in the organization, it is integrated to the basic functions (K1).
- K3: Finally, this variable is defined as a relation impact-benefit generated by the knowledge acquired in K2, in order to increase the staff's capacity to

satisfy the customers' needs and, as consequence, increase the competitiveness.

Therefore, K2 is a continuous cycle applied to K1 to obtain K3 and, consequently, improve the worker's process in order to obtain a better product or service and increase the customer satisfaction. Furthermore, the main strength of this model lies on the openness that provides to employees to contribute improvement ideas identified throughout their experience performing its basic functions. Otherwise, the continuous improvement remains in the hands of the employees who know and use the tools, while the rest of the staff only fulfill their activities. In fact, this work philosophy implies that the improvements can be identified by anyone who is involved in the processes leading to cultural changes within organizations. This idea is closely related to SPI because it assumes that a need for improvement may come from top management or managerial and operational levels, due to their direct contact with the process. Another more particular advantage of the model establishes the necessity to work from the basis, something like the analogy of the bullet train where the high priority is given to build the basic rail network, and posteriorly the bullet train is built. Thus, the Ki Wo Tsukau® model first builds and develops the organizational culture, thinking, and behavior by collaborating and exchanging ideas; subsequently it transfers these ideas into concrete actions in order to provide support to the continuous improvement cycle.

The model promotes the work by defining five basic tools, formally known as the 5 S, with the aim to improve the quality of life. These tools represent a quality practice devised in Japan related to the "integral maintenance" of an organization. This approach, known in the United States as "housekeeping", moves the concept of internal management of the home to work. This technique is called 5 S because it represents actions that are principles expressed with five Japanese words that begin with the S: *seiri* (classification), *seiton* (order), *seiso* (cleaning), *seiketsu* (standardization), and *shitsuke* (discipline and commitment). In the context of the software industry, the 5 S technique is one of the cornerstones of Lean Startup [15], that has been highlighted by addressing the launch of businesses and products through validated learning, scientific experimentation, the iteration between releases to shorten the product development cycles, and progress measurement to obtain valuable feedback from customers. Moreover, it is well known the application of Lean Startup in the context of software enterprises [16-18]. Nevertheless, the implementation of the Ki Wo Tsukau® model has not been explored in the context of SPI initiatives.

3 Correctly using the energy in SPI

Nowadays, software enterprises recognize the strong relevance of quality in their products by introducing some activities of the quality process models into their daily work. In the context of Software Engineering, this gradual change can be done through SPI. According to Herranz, Colomo-Palacios, de Amescua Seco, and Yilmaz [19], for any SPI initiative to succeed human factors, in particular, motivation and commitment of the people involved should be kept in mind. In spite of this, a high degree of failures in the SPI initiatives is still directly related to the lack of commitment and motivation. In this regard, commitment is commonly established and maintained by the first stage of an improvement model, such as IDEAL [20], and motivational aspects are frequently supported by communication through meetings, workshops, and discussion. However, small-sized software enterprises still face many problems when committing and motivating for improvement initiatives. Therefore, understanding how to successfully deal with these human aspects is one of the main challenges in SPI programs because these can negatively impact the effectiveness of the process improvement within small-sized software enterprises [21, 22].

In this context, in order to simplify the complexity of correctly initiate a SPI initiative in a small-sized software enterprise, we have designed a strategy based on the stages for implementing the Ki Wo Tsukau® model. This strategy aims to easily establish a strengthened commitment with the SPI initiative through the involvement and contribution of all participants. Basically, all unnecessary effort is eliminated by focusing on 6 Ki Wo Tsukau® steps (6 KWT steps): approbation, planning, formation/training, ideas generation, evaluation, and documenting (see Fig. 2). Thus, through K2-K1 cycles the commitment is strengthened by promoting the basic functions of each participant and supporting the employees' proactivity through workshops and discussion. Once that the K2-K1 cycles are generated a solid basis to begin the SPI initiative, in K3 the impact-benefit is analyzed to achieve the satisfaction of all participants (internal and external). In this regard, Table 1 summarizes the strategy designed to strengthen the SPI commitment.

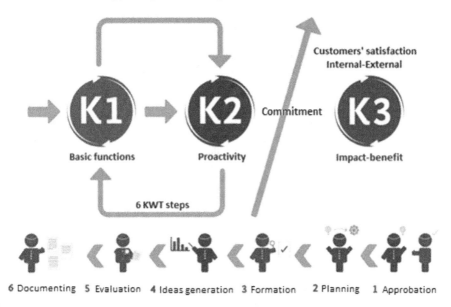

Fig. 2. Basic principles of the Ki Wo Tsukau® applied to strengthen the SPI commitment

Table 1. Strategy designed to implement the Ki Wo Tsukau® model into SPI initiatives.

Stage of Ki Wo Tsukau® model	Recommendation for implementing the model into SPI initiatives	Participants
Approbation	The reasons for adopting a process model and the expected impact must be identified. Moreover, the internal and external factors that may affect the success of the SPI initiative must be recognized (e.g., work environment, budget, participants, organizational politics). In addition, the commitment must be formalized in order to clarify the responsibilities and scope on generated ideas in the K2-K1 cycles.	Top management, consultant (internal or external)
Planning	The Ki Wo Tsukau® committee is composed by the top management (or a representative), project managers (if there is more than one), and software team. Taking into account the size of the organization, a Working Group and the Assessment Team are integrated with the same employees. Finally, a schedule is created for the improvement initiative.	Top management, project manager, software team, consultant (internal or external)
Formation/Training	The Ki Wo Tsukau® is trained in both the managerial and operational approaches of the process model. The Assessment Team should also be trained in the evaluation technique following the Ki Wo Tsukau® model to keep the environment clean and ordered. This recommendation promotes the ideas generation and the K2-K1 cycles through early formation.	Top management, project manager, software team, consultant (internal or external)
Ideas generation	As consequence of the formation, it is highly recommendable that the software team provides insight about what technical knowledge would be necessary to receive before starting the initiative.	Top management, project manager, software team, consultant (internal or external)
Evaluation	Finally, all the previous information is recorded and reviewed for the participants in order to clarify any misunderstanding and to formally begin the initiative with the diagnostic of processes.	Top management, working group consultant (internal or external)

Furthermore, the step six is carried out once that each K2-K1 cycle is executed by documenting all the information generated in the five previous steps. Thus, a template was designed to help small-sized software enterprises in this task. This template is a MS Office Word macro document that provides helpful information to the enterprise and leads it to correctly document the achieved commitments. This additional support prevents the company is distracted by the creation of a documented commitment to serve not only to start the SPI initiative but to be useful throughout all improvement activities. All the created templates are totally free for any enterprise.

Moreover, taking into account the recommendations of the Ki Wo Tsukau® model, a cleaned and ordered environment has to be prepared to execute the 6 KWT steps in order to obtain proper levels of motivation. Additionally, with the aim to support this strategy, we have designed a training package that any company can use to strengthen commitment and motivation and clearly understand the scope and impact of an improvement initiative (see Fig. 3). This training material (PowerPoint slides) addresses three main topics which are focused on the context of a small-sized software enterprise: the top management commitment in SPI, the software team commitment in SPI, and the SPI benefits.

The main idea is to provide useful guidelines to small-sized software enterprises when they are planning to begin a SPI initiative avoiding, at the same time, unnecessary and complex tasks which could demand an overwhelming effort. In other words, we aim to correctly use the energy of small-sized software enterprises when establishing the SPI commitment.

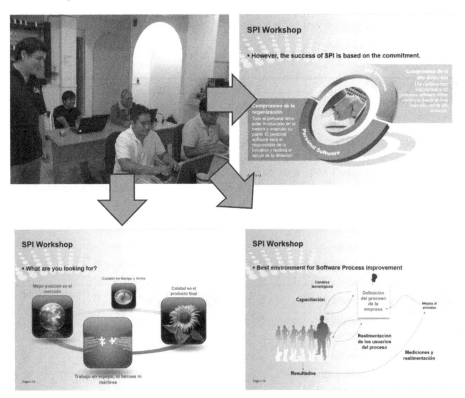

Fig. 3. Training package to support the commitment establishment

Finally, in order to validate the feasibility of this approach, we have conducted a preliminary validation in a Mexican small-sized software enterprise. The following section provides the details.

4 Preliminary results

An early validation was conducted in a small-sized software enterprise with 13 employees. This enterprise is located in Mexico City since 2010 and it is focused on developing accounting software and diverse applications for payroll control. An important aspect in relation to the choice of the participating enterprise was its disposition to cooperate for executing the previously defined strategy. It is noteworthy that this has facilitated the establishment of a serious commitment, coupled with adequate provision of resources in terms of personnel and facilities.

Furthermore, we decided to separate the staff into two groups of six members: the control group and the experimental group. The only member of top management was integrated in both groups. The control group followed the MoProSoft® guidelines, as process model, without any additional support; while the experimental group implemented the strategy that incorporates the principles of the Ki Wo Tsukau® model. Due to space reasons, the specific details for this validation were not included, but only the information necessary for presenting the obtained results. In this regard, Table 2 shows that the required time only for establishing and understanding the SPI commitment in the small-sized software enterprise was measured.

Table 2. Comparison of time used by the control and experimental groups.

Phase of the SPI initiative	Control group	Experimental group
Initiating (commitment establishment)	6 days	2.5 days

Additionally, we designed a questionnaire for obtaining a qualitative measurement on both groups by using a 4-point Likert scale wit "Strongly disagree" (SD), "Disagree" (D), "Agree" (A), and "Strongly agree" (SA) as responses. This 4-point scale was intentionally selected to eliminate the factor of indecision among employees using "socially desirable elements" as a semi-forced measure to allow employees to choose one side (positive or negative). Table 3 summarizes the numbers of responses provided by both groups participating in the validation of our approach.

Table 3. Responses obtained in the validation of our approach.

Statement	Control group				Experimental group			
	SD	D	A	SA	SD	D	A	SA
We have received the proper training to understand the SPI initiative	7	5					1	5
We have understood the SPI commitment by working as a team		6					4	2
We use clear templates to define the SPI commitment		6					1	5
We were continuously motivated during our participation in workshops		6					5	1
We have documented all our concerns and we have clarified them in discussions	5	1						6

It is clear that these results are very promising but these are not useful to generalize the benefits of this approach; nevertheless, we are currently designing one case study to support this approach by providing more quantitative evidence.

5 Conclusions

The preliminary validation enables to get some lessons related to the introduction of the Ki Wo Tsukau® model for simplifying the start of a SPI initiative by strengthening three basic principles for any improvement:

- Motivation: It is necessary to convince employees that, to working less, it is necessary to improve the working environment by eliminating problems caused by disorder, lack of technical knowledge, struggle of egos, etc. In this way, employees could correctly use their energy by "worry for" (or Ki Wo Tsukau) of solving the problems that currently exist in order to prevent their future repetition.

- Commitment: The elimination of the organization's problems reduces the losses related to product quality, delivery time, and development costs. In this regard, it is important to be clear about the commitment of each participant in order to maintain stable the working environment without affecting staff morale. For example, a bad impression that top management does not pay attention to their responsibilities within the improvement initiative can seriously affect the software team's motivation and wane, as a consequence, their energy. This can make that any improvement initiative fails before starting.

- Formation/Training: Attention must be paid to improve the standardization and discipline in carrying out the activities recommended in the process model by redesigning the procedures and templates that can facilitate their understanding. In this regard, it is necessary to involve employees in the development of procedures and templates.

References

1. Calero, C., Moraga, M. A., Bertoa, M. F., Duboc, L.: Green software and software quality. In: Calero, C., Piattini, M. (eds.), Green in Software Engineering, pp. 231-260, Springer International Publishing (2015).
2. Lesser, E., Ban, L.: How leading companies practice software development and delivery to achieve a competitive edge. Strategy & Leadership. 44(1), 41-47 (2016).
3. Laforet, S.: Organizational innovation outcomes in SMEs: Effects of age, size, and sector. Journal of World Business. 48(4), 490-502 (2013).
4. Srivastava, M. K., Gnyawali, D. R., Hatfield, D. E.: Behavioral implications of absorptive capacity: The role of technological effort and technological capability in leveraging alliance network technological resources. Technological Forecasting and Social Change, 92, 346-358 (2015).
5. McIntyre, S. G.: Learning lessons for organizational learning, process improvement, and innovation. In: McIntyre, S. G., Dalkir, K., Paul, P., Kitimbo, I. C. (eds.), Utilizing evidence-based lessons learned for enhanced organizational innovation and change, pp. 24-40, Hershey, PA: IGI Global (2014).
6. Psomas, E. L., Fotopoulos, C. V., Kafetzopoulos, D. P.: Critical factors for effective implementation of ISO 9001 in SME service companies. Managing Service Quality: An International Journal. 20(5), 440-457 (2010).
7. Ñaupac, V., Arisaca, R., Dávila, A.: Software process improvement and certification of a small company using the NTP 291 100 (MoProSoft). In: Dieste, O., Jedlitschka, A., Juristo,

N. (eds.), Product-Focused Software Process Improvement, pp. 32-43, Springer Berlin Heidelberg (2012).

8. Muñoz, M., Gasca, G., Valtierra, C.: Caracterizando las necesidades de las Pymes para implementar mejoras de procesos software: Una comparativa entre la teoría y la realidad. Revista Ibérica de Sistemas y Tecnologías de Información. 1(3), 1-15 (2014).

9. O'Connor, R., Basri, S., Coleman, G.: Exploring managerial commitment towards SPI in small and very small enterprises. In: Riel et al (eds.), Systems, Software and Services Process Improvement, pp. 268-278, Springer Berlin Heidelberg (2010)

10. Clarke, P., O'Connor, R.: An empirical examination of the extent of software process improvement in software SMEs. Journal of Software: Evolution and Process. 25(9), 981-998 (2013).

11. Dutra, E., Santos, G.: Software process improvement implementation risks: A qualitative study based on software development maturity models implementations in Brazil. In: Abrahamsson, P., Corral, L., Oivo, M., Russo, B. (eds.), Product-Focused Software Process Improvement, pp. 43-60, Springer Berlin Heidelberg (2015).

12. International Organization for Standardization.: ISO 9001:2008: Quality management systems. Geneva (2008).

13. Deming, W. E.: Out of the crisis. MIT Center for Advanced Engineering Study, MIT Press, Cambridge, MA (1986).

14. Kasuga, A., Verde, A.: "Modelo "Ki Wo Tsukau®, Preocuparse por... Aseguramiento de la mejora continua" Material de trabajo para docentes, México (2013).

15. Ries, E.: The lean startup: How constant innovation creates radically successful businesses. Portfolio Penguin: London (2011).

16. Wang, X., Conboy, K., Cawley, O.: "Leagile" software development: An experience report analysis of the application of lean approaches in agile software development. Journal of Systems and Software. 85(6), 1287-1299 (2012).

17. Bosch, J., Olsson, H., Björk, J., Ljungblad, J.: The early stage software startup development model: A framework for operationalizing lean principles in software startups. In: Fitzgerald, B., Conboy, K., Power, K., Valerdi, R., Morgan, L., Stol, K. J. (eds.) Lean Enterprise Software and Systems, vol. 99, pp. 1-15, Springer Berlin Heidelberg (2013).

18. Giardino, C., Unterkalmsteiner, M., Paternoster, N., Gorschek, T., Abrahamsson, P.: What do we know about software development in startups? IEEE Software, 31(5), 28-32 (2014).

19. Herranz, E., Colomo-Palacios, R., de Amescua Seco, A., Yilmaz, M.: Gamification as a disruptive factor in software process improvement initiatives. Journal of Universal Computer Science, 20(6), 885-906 (2014).

20. McFeeley, B.: IDEAL: A user's guide for software process improvement. CMU/SEI-96-HB-001. Software Engineering Institute, Carnegie Mellon University, Pittsburgh, PA (1996).

21. Sulayman, M., Urquhart, C., Mendes, E., Seidel, S.: Software process improvement success factors for small and medium Web companies: A qualitative study. Information and Software Technology, 54(5), 479-500 (2012).

22. Viana, D., Conte, T., Vilela, D., de Souza, C. R. B., Santos, G., Prikladnicki, R.: The influence of human aspects on software process improvement: Qualitative research findings and comparison to previous studies. In: 16th International Conference on Evaluation & Assessment in Software Engineering (EASE 2012), pp. 121-125. IEEE Computer Society (2012).

Analysis about the implementation level of ITIL in SMEs

Rubio-Sánchez, J.L[1], Arcilla-Cobián, M.[2], and San Feliu, T.[3]

[1] Universidad a Distancia de Madrid, Escuela de Ciencias Técnicas e Ingeniería, Carretera de La Coruña, Km 38,500, Via de Servicio 15, 28400 Collado Villalba, España
juanluis.rubio@udima.es
[2] Universidad Nacional de Educación a Distancia, ETS de Ingeniería Informática, 28040 Madrid, España
marcilla@issi.uned.es
[3] Universidad Politécnica de Madrid, ETS de Ingenieros Informáticos, 28660 Boadilla del Monte, Madrid, España
tomas.sanfeliu@upm.es

Abstract. Information Technology and Infrastructure Library (ITIL) is a fundamental need for any company or industry, especially for Small and Medium Enterprises (SMEs). However, there is an increasing demand on how to proceed to correctly implement it. There is not a clear guide, methodology or algorithm for it. The first step to take into account is to understand the current implementation level of ITIL processes in SMEs. Knowing the current level will help to set up a solution. In this paper, we present the results and analysis of a survey done in 64 small and medium enterprises of Madrid, Spain. The results let us knowing the current implementation level of ITIL processes.

Keywords: ITIL; SME; processes; implementation level.

1 Introduction

Due to the growth of software companies, a lot of new software tools for business were developed. Many large, medium and small companies have adopted Information Technologies (IT) based solutions. On this scenario, new services (based on IT) and new business models have emerged. The new companies demand new software and more services to technological companies as pointed in Mas et al. [1]. ITIL (Information Tecnology Infrastructure Library) is one of the standards that companies follow when implementing IT services. ITIL is used in large companies, because they have resources dedicated to implement it.

Analysing official data from the Ministry of Industry in Spain [2], around 40% of companies have less than 10 employees. And the data are even more significant if we take into account that almost 95% of companies are small and medium enterprises (less than 250 employees) in Spain.

The small companies have not specific resources dedicated to implement ITIL. It is even surprising the percentage of companies operating their business far from the ITIL processes, as pointed in Network World [3]. This is something identified in Binders et al. [4] and Shang et al. [5]: the fact is that ITIL is long, expensive and risky

© Springer International Publishing AG 2017
J. Mejia et al. (eds.), *Trends and Applications in Software Engineering*, Advances in Intelligent Systems and Computing 537, DOI 10.1007/978-3-319-48523-2_2

for SMEs. Anyway it is important to highlight that most of the procedures described to implement ITIL just consider factors that may affect its implementation and how to deal with them. Some examples can be found at Ahmad et al. [6] and Pollard et al. [7].

This leads to the question: Is it possible these small companies follow ITIL? i.e., with such few resources, is it really possible to implement the processes recommended by ITIL? Should it be needed to adapt ITIL to SMEs? What are SMEs doing today related to the implementation of ITIL?

Taking in consideration what has been previously exposed, we are looking for how to help SMEs during the implementation of ITIL. The main objective of this paper is to know the current situation of SMEs in terms of ITIL implementation. In this way, a survey has been conducted to know how SMEs deal with processes, standards, certifications and specifically which experiences they have related to the processes of ITIL. The specific objectives of this research related to SMEs are:

- to know which ITIL processes have been widely implemented in SMEs.
- to gather information about the most interesting ITIL processes for SMEs (some processes could be interesting for a SME but have not been implemented for different reasons, for example its high implementation cost).
- to analyse which ITIL processes will be implemented in short term in SMEs.

The rest of the sections are organized as follows: initially an overview is made to get information about ITIL and SMEs. After that, the technical aspects of the survey are shown. We continue analysing the results of the survey and, finally, some conclusions are discussed.

2   Overview of ITIL and SMEs

2.1 Information Technology Infrastructure Library

ITIL is the most common approach to ITSM (Information Technology Service Management). In fact, ITIL is a de facto standard -as pointed in Hochstein et al. [8] to provide quality services. The current ITIL practices are the result of multiple experiences and several reviews of its previous versions. The current version is ITILv3, updated in 2011 (known as ITIL 2011), and it was organized in five books: Service Strategy [9], Service Design [10], Service Transition [11], Service Operation [12] and Continual Service Improvement [13].

The Service Strategy is oriented to design and implement the service from the perspective of capability and strategy. It defines a guided overview on the principles to be used for developing the policies and processes across the ITIL service lifecycle. The processes included in ITIL 2011 are: Strategy Management for IT Services, Service Portfolio Management, Financial Management for IT Services, Demand Management and Business Relationship Management, although companies are not really aware of them, as remarked in the survey of itSMF [14].

The Service Design is fully oriented to designing and implementing a service. The objective is to turn strategies and objectives into services useful to the company. It takes care for all the lifecycle of a service, the continuity of the service and the quality of the service. The processes included in ITIL 2011 are: Design Coordination, Service Catalogue Management, Service Level Management, Availability Management, Capacity Management, IT Service Continuity Management, Information Security Management and Supplier management.

The Service Transition processes are addressed to move services into operation. The main idea of these processes is to control risks through Transition Planning and Support, Change Management, Service Asset and Configuration Management, Release and Deployment Management, Service Validation and Testing, Change Evaluation and Knowledge Management.

There is another group of important processes described in ITIL practices orientated to Service Operation. These processes ensure efficiency when operating a service as well as the value for the client. The processes included are: Event Management, Incident Management, Problem Management, Request Fulfilment and Access Management.

Finally, there is a Continual Service Improvement process that describes the 7 step of the improvement process.

2.2 SMEs in Spain

It is well known that the number of large enterprises is really small compared to the number of SMEs. This is a typical distribution in many countries. Regardless the ratio of large/small companies differs from one country to another, the percentage of SMEs is over 90%. For example, taking data from latest years in one of the most industrialized regions in Spain (Madrid), the number of companies with less than 10 employees is around 88% and the number of companies with more than 250 employees is lower than 1%.

IT expenses are not uniformly distributed as pointed in Marquez et al. [15]. This leads to a situation where some regions with companies offering high value services need to implement ITIL processes to compete while other regions are just offering primary industry products and services. As indicated in Gorriti et al. [16], European countries have not taken profit of the introduction of IT in SMEs. Even if high investments have been made in technology, the lack of processes make the investments useless. There is no knowledge enough in SMEs about the world of practices, regulations, certificates and so. Even if there are some attempts to clarify it, as Muñoz et al. indicate [17], SMEs are not aware of them. SMEs are not worried about quality and processes, govern and IT management as indicated in Aragon et al. [18]. The introduction of IT in SMEs is different from one industry to another. For example, financial, media or technology industries innovate and adopt IT and IT regulations very quickly as pointed by Mas et al. [1] while manufacturing industry is slower.

3 Survey Methodology

In order to know the current implementation level of ITIL in SMEs a web survey was designed. A web questionnaire has been created and it was hosted in a public site, so all the answers could be saved on-line and analysed.

3.1 Scope

The scope of the survey was defined to companies operating in Madrid region. Madrid was selected because is the region in Spain with the highest number of companies. Initially, it was used a database containing around 10,000 companies. A subset of 150 companies was randomly selected. The companies included in this study were from different industries such as Education, Tecnology, Financial, Marketing and Logistics among others. A message was sent to invite them to participate.

3.2 The Survey

The companies were initially asked about general information: company name, number of employees, number of IT employees, age of the company, industry. After that, the companies were asked about their interests on practices for IT management (quality, risks, business, certifications, competitiveness and costs), results of the implementation of practices (control of services, improvement on ROI, better alignment of IT department and business, better IT governance, savings), knowledge of the SME about standards (ISO 20000, ISO 9000, ISO 27001, BS 25999, ITIL, COBIT, PMBOK, PRINCE2 and others).

And finally we proposed a set of questions specifically dedicated to the processes of ITIL. We asked about every specific process in ITIL. In this case, questions with three options for the answer were offered:

1. Not implemented and not planned.
2. Not implemented but planned at medium term.
3. Implemented or will be implemented in short term.

3.3 Technical Issues

The sampling company selection method was randomized and one staged. Once the period to answer has overcome, the total number of responses was 64. The results obtained from the survey indicate a confidence level of 90% and a sampling error not higher than 8.5%. The percentages of error must be interpreted, as the response is in the interval 2 ± 0.17, which still avoids overlapping with the rest of the intervals for responses 1 and 3.

4   Results and Analysis of the Survey

In this section we will present the main results of the survey conducted and a brief analysis and discussion about them. Most of the companies know about ITIL v3 but do not really know about the new processes of ITIL v3 2011 (in fact, it represents just a slight update), so we will present results related to the processes of ITIL v3.

4.1 Size

Approximately, 38% of the respondent companies have 1-9 employees, 45% have 10-49 employees and 17% have +50 employees. There is no large company in the survey and more than 80% of the companies have 49 or less employees. This is coherent with the official results published by the Ministry of Industry in Spain [2] that suggested that 40% of Spanish companies have 9 or less employees.

Anyway, the results in Fig. 1 are much more significant. In this case we have represented the answers classified by the number of employees in the company. The results show the bigger the companies are, the better the results are. That is, as the staff is bigger, the company seems to be more prepared to implement ITIL processes. That was really the interest of the study, as mentioned in the introduction section. The question we arose was if the SMEs could implement these ITIL processes with such small staff and facts show the hypothesis was correct. This result pushes us to help SMEs to implement ITIL processes.

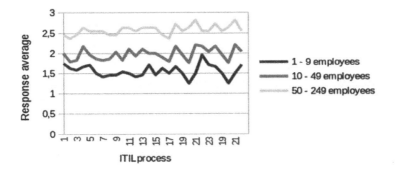

Fig. 1. Number of responses by size

4.2 Strategy processes

The results for the implementation level of ITIL strategy processes are shown in Fig. 2. The most relevant issue in this point is that the average for the three processes included in the survey, take a value close to 2. This could mean that SMEs have not implemented these processes but they have planned to do it in a medium term. As we will comment later this is dependent on the value of variance because depending on its value the results could mean that SMEs are planning to implement the strategy group of processes in a short term or it could also mean (if variance is high) that some SMEs have already implemented these processes and some others have not implemented them (and they do not have planned to do it).

Fig. 2. Coverage of the ITIL Service   *Fig. 3. Coverage of the ITIL Service Design*
Strategy processes   *processes*

One of the questions in the survey was related to the implementation level of the service strategy processes. In fact, the companies were asked if they thought they had implemented processes for managing the strategy. In most of the cases the response was negative. But when the companies were asked for every independent process, their answers were affirmative, i.e., they had already implemented those processes or they will have them in a short term.

This situation may be caused by the fact that companies do not realize they are implementing ITIL processes when they really are. This obeys to the need to implement these processes because of demands of its own daily operation.

Nevertheless, there is another issue to take in consideration. Although what we have previously explained is one of the reasons, we must have a look at the variance of the responses. If we do, we will realize that the big variance obtained is appointing that companies usually have or have not implemented these processes but they do not usually have plans to do it. It may happen probably because these processes are not complicated to be implemented, that is, just when a SME decides to implement it, it can do it quickly with no long plans for it.

Finally, the conclusion about these processes is that it is not easy to find companies that have implemented the service strategy processes. We must think how far a SME is from managing the financial aspects of its IT department or how far is from planning the demand. Of course they will always prefer to assign resources to daily operations than managing demand or financial aspects of its IT area.

4.3 Design processes

As we can see in Fig. 3, SMEs are far from having these processes implemented. Just taking a look at the results, SMEs do not even have planned to implement some of the processes. In this case, it must be appointed a great variance in results, what represents big differences among all companies analysed.

We must appoint that the specific process referred to the service catalogue management is not implemented in many cases and will not be in a short future. This is a relevant and worrying result because it means that IT departments do not always have a clear service catalogue which turns in accepting almost any kind of job.

We must pay also special attention to continuity and security management processes as the average is lower than 2. This means that companies do not really care about both topics. That is something we could expect, as these processes are expensive and not easy to be implemented.

The results let us think there is no a clear activity to manage the suppliers, even if we could think this is a basic process. If we think in any company, but especially on those SMEs that hardly have resources, this should be one of the processes most assumed. But the fact it is that the situation is far from this. The situation is that due to the small staff (and so small size) of the companies, SMEs cannot dedicate resources to follow a supplier management process.

The results for the capacity process and availability of resources are better than the rest of the service design processes, which means that companies try to manage their capability and resources. This is happening because they represent direct costs and so SMEs are more sensitive to the management of these processes.

4.4 Transition processes

The results are really polarized in extremes again, what leads to a really significant variance. If we just get conclusions on the average values, we could think that SMEs have not implemented transition processes but they will in a short future. But that is not the case. An average close to a value of 2 – as we can see in Fig. 4, is meaning either companies are thinking about a medium term implementation of transition processes or -and that is the case- there is a great variance between those companies interested on ITIL processes and another ones that are not.

This group of processes is critical for any IT department or ICT based company. This explains why the planning and delivering processes are better scored than the rest of the processes, even compared to the processes of the other groups. We must think that these companies or IT departments are forced to deliver results.

Anyway, we must pay attention to the fact that some processes are not implemented in most of SMEs, for example, the knowledge management process.

Something similar can be concluded for the availability management process. The reason is that SMEs are just focused on delivering in any manner, no matter the conditions.

The conclusion with the validation process is not different: being as relevant as it is for the quality of a service, it is surprising that all SMEs do not have it fully implemented.

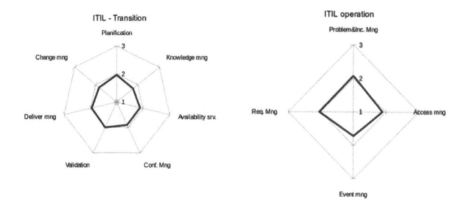

Fig. 4. Coverage of the ITIL Service   Fig. 5. Coverage of the ITIL Service
           Transition processes   Operation processes

4.5 Operation processes

Regarding the group of processes for the service operation, some issues were grouped in the same question just to avoid a longer questionnaire. It is relevant the fact that incidents and problems really matter and SMEs are more sensitive to them than to other ITIL processes. If we take a look to Fig. 5, we can see that results are also really close to a value of 2 (in the case of problem and incident management processes is a bit higher) but in this case the variance obtained is lower which indicates how SMEs take care about these processes.

Nevertheless, it cannot be concluded that this is a solved question for SMEs because the averages obtained are not at the top but still far from it. It is a fact that results are slightly better than in other groups of processes but it is possible to find improvement actions for SMEs. Even though SMEs do not clearly establish differences between some processes, for example *Problem Management* and *Incident Management*. This topic is aligned with the response obtained in one of survey questions that demanded information about the SMEs knowledge on ITIL and similar practices. This issue can be also determined from the responses obtained when asked about different references about good practices. SMEs do not really see any difference among ITIL, COBIT, COSO and others.

If we take a look at the results of request fulfilment, incident and problem management, a high percentage of companies have implemented the associated processes. Some other processes as event management or access management have not been implemented. What can be concluded again is that IT departments have just implemented the processes they 'consider' really relevant for the business. We highlight 'consider' because, in this case, there is no reason for not having implemented the event management process or the access process.

Unlike it happened with the security and continuity processes, the event management does not represent an extra cost as it can be implemented with the same resources than the incidences resources.

Something similar can be said about the access management process. A false security feeling drives SMEs to not implementing the access management process. As it can be deduced from the survey, small companies do not pay attention to access management. This may obey to the fact that security is not a tangible asset and so the perception of risk is not so high.

4.6 Continual Service Improvement process

Finally, the coverage of the continual service improvement process was 2.2. In this case there is a result better than in the other group of processes. This may be caused by well-known standards and practices such as CMMI (in any of its representations), Six Sigma or LeanIT, as SMEs seem to be really aware of its importance.

There were some questions about the reason to implement the practices related to this process, and the SMEs answered that the main reasons were cost savings, quality level improvement, business needs and competitiveness improvement. So, even there is not a unique objective when implementing these practices, SMEs identify the need to do it.

5 Conclusions

The main conclusions we have obtained up to now are summarized in the next points:

1. Only very few studies are focused on SMEs. In general, studies do not distinguish between large and small companies.
2. The results show us that no process stands out above other. In any case a guide to help SMEs to take a decision on the process to be implemented is necessary.
3. The data obtained in the survey show that companies are very polarized: either they have already implemented (or they are on the path) most of ITIL processes, either they have not started yet.

This paper has provided an insight into the current business practices in Spain, in particularly the SMEs related to the region of Madrid.

References

1. Mas, M., Quesada, J.: Las nuevas tecnologías y el crecimiento económico en España. Fundación BBVA (2005)
2. Dirección General de Industria y de la Pequeña y Mediana Empresa: Retrato de las PYME 2013. Ministerio de Industria de España (2013)
3. NetWorkWorld, http://www.networkworld.es/actualidad/un-64-de-las-empresas-ve-en-itil-la-clave-para-mejorar-la-reputacion-de-las-ti

4. Binders Z., Romanovs A.: ITIL Self-assessment approach for small and medium digital agencies. Information Technology and Management Science, 17, 1, 138-143, De Gruyter Open (2014)
5. Shang, S.S.C. and Lin, S.F.: Barriers to implementing ITIL – a multi-case study on the service-based industry. Contemporary Management Research, Vol. 6 No. 1, 53-70 (2010)
6. Ahmad N., Tarek N., Qutaifan F., Alhilali A.: Technology adoption model and a road map to successful implementation of ITIL. Journal of Enterprise Information Management 26:5, 553-576 (2013)
7. Pollard, C.: Justifications, strategies, and critical success factors in successful ITIL implementations in US and Australian companies: an exploratory study. Information Systems Management, Vol. 26 No. 2, 164-175 (2009)
8. Hochstein A., Zarnekov R., Brenner W.: Evaluation of service oriented IT management in practice. In: Proceedings of ICSSSM'05 2005 International Conference on Services Systems and Services Management, pp. 80-84 Vol. 1 (2005)
9. Cabinet Office: ITIL® Service Strategy, 2nd ed. Norwich, UK: The Stationery Office (TSO) (2011)
10. Cabinet Office: ITIL® Service Design, 2nd ed. Norwich, UK: The Stationery Office (TSO) (2011)
11. Cabinet Office: ITIL® Service Transition, 2nd ed. Norwich, UK: The Stationery Office (TSO) (2011)
12. Cabinet Office: ITIL® Service Operation, 2nd ed. Norwich, UK: The Stationery Office (TSO) (2011)
13. Cabinet Office: ITIL® Continual Service Improvement, 2nd ed. Norwich, UK: The Stationery Office (TSO) (2011)
14. Institute of Systems Science: itSMF 2013 Global Survey on IT Service Management. itSMF International (2013)
15. Márquez Ramos, L., Martínez Zarzoso, I., Sanjuan Lucas, E., Suárez Burguet, C.: Efecto de las TIC sobre el comercio y el desarrollo económico. Análisis para el caso de España. Estudios de Economía Aplicada, 25 (1) (2007)
16. Gorriti, M., Álvarez, J. L. R.: La contribución de las TIC al crecimiento económico en España y los retos del sector. Presupuesto y gasto público 39, 243-266 (2005)
17. Muñoz Periñán, I.L., Ulloa Villegas, G.V.: Gobierno de TI-Estado del arte. Sistemas y Telemática (2011)
18. Aragón Sánchez, A., Rubio Bañón, A.: Factores asociados con el éxito competitivo de las pyme industriales en España. Universia Business Review 8, 38-51 (2005)

Sustainability in Software Engineering - A Systematic Mapping

Kristina Rakneberg Berntsen, Morten Rismo Olsen, Narayan Limbu, An Thien Tran and
Ricardo Colomo-Palacios
{Kristina.r.berntsen, morten.r.olsen, narayan.limbu, an.t.thien, ricardo.colomo-
palacios}@hiof.no

Abstract: Information Technology (IT) has become a key element in our everyday life, and one of humanity's current challenges is to conserve the environment and attain a sustainable IT development. Therefore, it has become increasingly important how environmentally friendly a software product is during its life cycle and the effects on the environment related to the development, exercise, maintenance and disposal of the software product. The purpose of this study is to outline recent development of frameworks and guidelines in sustainable software engineering. A systematic mapping was conducted which focuses on practices and models that are being used or proposed in this regard. The results reveal different types of models and different criteria for evaluating sustainability properties. In addition, the study indicates an increase of interest in this field in recent years whereas results suggest a handful of prominent authors and venues publishing research within the scope of sustainable software engineering.

Keywords: Sustainable software engineering, systematic mapping.

1 Introduction

IT is used frequently in today's business world and is an essential part of most business strategies. However, the impact of IT has a significant negative impact on the environment. Ehrlich and Holdren [1] presented a function to determine the environmental impact of human consumption. It is estimated that the electricity consumption of the ICT sector will increase by nearly 60 percent from 2007 to 2020. This expresses a need for a more sustainable development (SD) in order to reduce energy consumption and greenhouse emissions (GHG).

In recent years, frameworks and guidelines for sustainable software engineering (SSE) have emerged. A study in 2014 by Ahmad et al. [2] explored the evolution of frameworks and guidelines in the SSE which concluded that "Sustainable development should integrate social, environmental, and economic sustainability and use these three to start to make development sustainable" [3]. There is a definitive trend of research regarding sustainable software engineering which has led to terms

© Springer International Publishing AG 2017   23
J. Mejia et al. (eds.), *Trends and Applications in Software Engineering*, Advances
in Intelligent Systems and Computing 537, DOI 10.1007/978-3-319-48523-2_3

such as Green IT. The focus on Green IT in recent years have shown that SSE is gaining more attention in the IT industry in terms of software products, but also concerning IT management.

Since the GREENSOFT model [4] was presented back in 2011, there has been an increased amount of models and guidelines for sustainable software development (SSD). In addition, there are several suitable definitions for sustainable software, however the authors chose the two following definitions of sustainable software:

Definition 1: "Sustainable software is software whose direct and indirect negative impacts on economy, society, human beings, and the environment resulting from development, and usage of the software is minimal and/or has positive effect on sustainable development" [5].

Definition 2: "The art of developing sustainable software with a sustainable software engineering process so that negative and positive impacts result in and/or are expected to result from the software product over its whole life cycle are continuously assessed, documented, and used for further optimization of the software product" [6].

Penzenstadler et al. [7] conducted a systematic mapping study which focuses on different research topics within SD. It argues that "Software Engineering for Sustainability (SE4S) has received widespread attention in the SE community over the past few years." The study concluded that SE4S was an immature area of research at the time. Their paper however shows clear indications that research within the fields of software engineering models and methods have gained momentum since 2010. Ahmad et al. [2] conducted a systematic literature review research on sustainability studies on software engineering, from which an overview of research activities, limitations, approaches and methods was developed. They came to the conclusion that sustainable development is an emerging challenge for software engineering and needs improvement based on the development process, management, and evaluation. The paper also reported that there were 13 articles discussing approaches and methods of sustainability in software engineering.

According to Ahmad et al. outlines existing systematic literature review publications on sustainable SE [2], [8]. Their emphasis however, was not primarily placed on models and guidelines in the scope of SSE. Thus, in this paper, a systematic mapping to outline proposed practices and explore the field of SD in SE was conducted. In addition, the scope covers the role of sustainable development in the IT industry. The next section will describe the methodology used for this purpose. Section 3 presents the results of the study, and section 4 presents the conclusion.

2 Research methodology

The focus of this study is to outline the "state of the art" of SSE and elaborate on existing and proposed models, guidelines and practices in this regard. Thus, the propulsive research question for this study is:

What guidelines and models exist in current research for SSE?

Considering the nature of the topic, a systematic mapping study was conducted in reference to the guidelines according to Petersen et al. [9]. In the sections that follow, the research questions (RQs) are described and the correlating metrics for this study is presented.

2.1 Research questions

In order to ensure that this aim is achieved, the authors have developed the following RQs:

- *RQ1:* What are the most cited/reported guidelines/models for SSE?
- *RQ2:* What is the evolution of interest in SSE?
- *RQ3:* What are the most important authors and venues on this topic?

2.2 Search strategy

In this section the search strategy for finding literature on the topic is presented. In order attain relevant data, the most popular academic databases in the field of information systems and computer science were used:
ACM Digital Library (http://dl.acm.org)

- IEEE Xplore Digital Library (http://ieeexplore.ieee.org)
- ScienceDirect (http://www.sciencedirect.com)
- Springer Link (http://link.springer.com)
- Wiley Online Library (http://onlinelibrary.wiley.com/)

Regarding the search string and keywords, the authors experimented with different combinations of search strings in different databases and jointly established them in the final search for papers:
("sustainability" OR "green*" OR "environment*") AND "software engineering" AND ("model" OR "framework" OR "specification")

2.3 Study selection

The main criterion when selecting papers was the focus on papers regarding SSD. Furthermore, the keyword "Green IT" with a focus on the key phrase "green", was a central metric for the study selection. The study was based on recent development in the field. Another metric was that the selected papers should be available through the Østfold University College library. In order to avoid researcher subjectivity, the authors unanimously chose the following exclusion criteria:

- *Based on accessibility*: Libraries that Østfold University College does not have access to.
- *Based on publication date*: Papers before 2010.
- *Based on language*: Papers that are not written in or translated into English.
- *Based on titles*: The title does not indicate any relation to SSD.
- *Based on abstract*: The abstract does not describe SSD.
- *Based on full text*: The content is not relevant to the RQs.

2.4 Study classification

In the study classification the papers that were selected from the initial inclusion phase were grouped according to their keywords. Additional criteria were: name of author, year of publication, and venue. The categories were adapted to fit RQ1 as presented in the results section. After removing duplicates and merging similar keywords, 32 keywords were used as a basis for the following 6 categories:

- *Energy Efficiency*: Papers that present energy consumption calculation methods, energy efficient systems, and/or metrics for optimizing system performance with lower costs.
- *Development Methodology*: Papers that present improved algorithms or frameworks for software development.
- *Process Enhancement*: Papers that focus on specific sub-section of processes used in the life cycle of a software product.
- *Organizational Metrics*: Papers that are directed towards administrations and present evaluation of current procedures or papers that aim to improve these.
- *Life-Cycle Thinking*: Papers that cover all processes in the software product life-span. Papers that evaluate several processes or aim to improve these.

2.5 Data extraction

This section presents the results from the study classification section. The main focus is placed upon IEEE which contains 19 papers. Out of a total of 105, only two authors appear twice, those being Naumann [4] and Dick et al. [10]. The distribution of publication dates shows that there is a majority of papers in the range of 2013 to 2015 with a peak in 2014, which is represented 10 times. Finally, there was arguably a majority of "solution proposal" papers in the selection.

3 Analysis and discussion of results

In this section the findings from the systematic mapping are presented (see Table 1). All selected papers have made it through a filtering process where the level of relevance to the RQs was determined. The data extraction requirements are described in section 3.3.

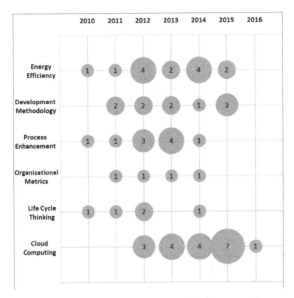

Fig. 1. Distribution of primary studies by categories.

After filtering the papers based on title, keywords, date and abstract, 99 papers were singled out for full-text reading. This resulted in 36 papers. Then they were categorized to answer the RQs, presented in section 3.1.

Table 1. Paper filtering phases

Library	Total number of hits	Abstract selected	Full text selected
ACM	2754	19	7
IEEE	168	35	17
ScienceDirect	7413	16	7
Springer	7712	7	1
Wiley	3802	21	4
Results	**20824**	**98**	**36**

The next section provides an overview of all the RQs with a discussion to determine if the articles comply with the objective of this study.

3.1 RQ1: What are the most cited/reported guidelines/models for SSE?

The authors extracted 20 articles related to models or guidelines used for SSE. In addition, the authors included papers regarding metrics for these guidelines to answer RQ1. The papers were grouped into 5 categories based on the most frequently occurring topics in the papers (see Table 2).

The literature points to the GREENSOFT model as an early suggested model for "green" and SSD. Early versions of this model have been traced back to 2010 33333[5]. A year later, the model was presented more elaborately. The GREENSOFT model is a conceptual reference model with criteria and metrics [4]. The other part of the model shows the life cycle for software products better known as "cradle-to-grave" coverage of software whole life cycle of software products.

- *Energy Efficiency*: Energy consumption model [11], Energy benchmarks and metrics assignment [12], Data Center Power Metrics [13].
- *Development Methodology*: Ripple effects considerations [4], Generic Sustainable Software Star Model (GS3M) [14].
- *Process Enhancement*: Tailored to Agile Methodology [10], Interaction Design Employment [15].
- *Organizational Metrics*: Holistic Green IT strategy [16], Green Performance indicators [17].
- *Life-Cycle Thinking*: "Cradle-to-Grave" - optimization [4], Life-cycle assessment [13].

Table 2. Distribution of topics in papers regarding RQ1.

Mapping	Energy Efficiency	Development Methodology	Process Enhancement	Organizational Metrics	Life Cycle Thinking
[17]	X				
[15]		X	X		
[10]			X		
[18]	X				
[19]		X	X		
[20]		X		X	
[14]		X		X	
[21]				X	
[12]	X		X		X
[22]	X				
[4]		X	X		X
[5]	X		X		X
[23]			X		X
[16]			X		X
[24]		X	X		
[25]		X	X	X	
[13]	X				X
[26]		X	X		X
SUM	**6**	**8**	**11**	**4**	**7**

According to the results, there is a growing attention on the aim for sustainability in every stage of software development. The impact of software on the environment is not solely due to the hardware or servers it makes use of. Davor Svetinovic [25] argues that requirement engineering is fundamental to the process of making complex sustainable systems. Additionally, Mahmoud et al. [17] proposed that testing phase, which was neglected in the GREENSOFT model, should be included in the software product life cycle since it can be environmentally detrimental as well. Thus it can be observed that researchers now regard the life cycle of software from a more comprehensive perspective.

3.2 RQ2: What is the evolution of interest in SSE?

Our findings from this systematic mapping shows that although sustainability is not supported by traditional software engineering methods [27], there exists a significant increase of interest for developing both reliable and long-lasting software with SSE methods [28]. As illustrated in Figure 4, there has been an increase in interest for SSE during the past four years. The graph represents the results from our systematic mapping where 36 primary studies were selected and the years on which they were published. Areas of interest varies from energy savings (green IT) and business processes to non-functional requirements that are aligned with sustainability principles [29]. In the last four years there has been an increased number of publications in sustainability for software engineering in various domains [2]. The most interesting publications are articles regarding SSE frameworks or models. Dick et al. [10] presents a model that integrates Green IT aspects into software engineering processes with agile methods in order to produce "greener" software from scratch. Burger et al. [30] formulated a methodological approach which aims to formulate adequacy conditions for concepts of sustainability. As well as illustrate a categorical framework with the required general concepts, and propose a conception of sustainability based on the capability approach.

There has been a significant increase in interest in SSE between 2011 and 2014. The graph shows a drastic decrease from 2015. Since the graph represents the results from our systematic mapping, the decrease of interest may be caused by our lack of findings of published papers in the year of 2015. Thus the graph may be lacking in representing the real-life situation when it comes to interest in SEE.

3.3 RQ3: What are the most important authors and venues on this topic?

Findings indicate two venues that distinguish from the others:

- The International Workshop on Green and Sustainable Software (GREENS) is an annual workshop focusing on green software engineering. The first workshop was in 2012. The goal has been to bring together academics and practitioners to exchange experiences and results of previous and ongoing practices and research on green and sustainable software.
- The Harnessing Green IT book is the other venue that produced several hits even in the preliminary studies. Two of which made it through the data extraction and by so, deemed as relevant literature regarding green IT and green computing. The book presents the principles, methods and solutions used in green IT practices.

Table 3. Distribution of topics in papers regarding RQ3.

Venues	Reference
International Conference on Software Engineering (ICSE)	1
Journal of Sustainable Computing: Informatics and Systems (SUSCOM)	2
International Conference of Chilean Computer Science Society (SCCC)	1
Harnessing Green IT: Principles and Practices Book	2
International Workshop on Green and Sustainable Software (GREENS)	3
Journal of Systems Engineering	1

Conference on Human Factors in Computing Systems (CHI)	1
ACM SIGSOFT Software Engineering Notes	1
Latin American Computing Conference (CLEI)	1
International Conference on Evaluation and Assessment in Software Engineering (EASE)	1
Journal of Environmental Impact Assessment Review	1
Conference on Advances in Computing and Communications (ICACC)	1
The 4th International Conference on Ambient Systems, Networks and Technologies (ANT)	1
IFIP Advances in Information and Communication 2010	1
Green Technologies Conference (GreenTech), 2015 Seventh Annual IEEE	1
Total	**19**

Regarding authors, Stefan Naumann, Markus Dick and Eva Kern contributed to the most papers [5], [4], [22], [10] whereas the last one was published in the GREENS workshop 2013. Birgit Penzenstadler is accredited the most unique papers in the data selection [31], [27], [7], [8]. Table 4 outlines an outtake of the distribution of authors contributing to papers in the extracted data. The table illustrates the collaboration of Eva Kern, Markus Dick and Stefan Naumann in several papers in this selection. The data extraction resulted in a variety in terms of authors with the given scope. There is circumstantial evidence of the aforementioned authors conducting research that is accepted in SSD venues (see Table 4). Furthermore, the contribution of Birgit Penzenstadler is mentioned due to a systematic mapping studies on software engineering for sustainability seeing as this contribution is relevant reference literature for further research.

Table 4. Author contributions in extracted literature.

Author	Contribution
Dick	[9], [10], [16], [28]
Kern	[9], [16], [28], [29]
Ardito	[18], [38]
Naumann	[9], [10], [28], [29]
Penzenstadler	[5], [12], [33]
Murugesan	[22], [32]

4 Conclusion

In this systematic mapping, current practices in sustainable software engineering are outlined. This paper focuses on models that are being used or proposed in this regard. Sustainable development is the emerging challenge for software engineering and needs improvement based on the development process, management and evaluation. Similar research has been done in the SSE field [2], [7]. This paper differs regarding the focus on models used in the SSE field. The results reveal different types of models and a different set of criteria for evaluating sustainability properties. The sustainability aspect is being taken into account in every step of software development.

In addition, there is a clear indication of increased interest in this field in recent years. Results also suggest a handful of prominent authors and venues publishing research within the scope of sustainable software engineering. The importance of SEE is prominent. However, our results may have been lacking due to the strict inclusion/exclusion criteria. As for future work, other databases may provide different results, thus it can be a motive for further exploration.

References

1. Holdren, J.P., Ehrlich, P.R.: Human Population and the Global Environment: Population growth, rising per capita material consumption, and disruptive technologies have made civilization a global ecological force. Am. Sci. 62, 282–292 (1974).
2. Ahmad, R., Baharom, F., Hussain, A.: A Systematic Literature Review on Sustainability Studies in Software Engineering. Knowl. Manag. Int. Conf. KMICe Langkawi Malays. (2014).
3. Goodland, R.: The Concept of Environmental Sustainability. Annu. Rev. Ecol. Syst. 26, 1–24 (1995).
4. Naumann, S., Dick, M., Kern, E., Johann, T.: The GREENSOFT Model: A reference model for green and sustainable software and its engineering. Sustain. Comput. Inform. Syst. 1, 294–304 (2011).
5. Dick, M., Naumann, S., Kuhn, N.: A Model and Selected Instances of Green and Sustainable Software. In: Berleur, J., Hercheui, M.D., and Hilty, L.M. (eds.) What Kind of Information Society? Governance, Virtuality, Surveillance, Sustainability, Resilience. pp. 248–259. Springer Berlin Heidelberg, Berlin, Heidelberg (2010).
6. Venters, C.C., Jay, C., Lau, L.M.S., Griffiths, M.K., Holmes, V., Ward, R.R., Austin, J., Dibsdale, C.E., Xu, J.: Software Sustainability: The Modern Tower of Babel, http://ceur-ws.org/Vol-1216/paper2.pdf.
7. Penzenstadler, B., Raturi, A., Richardson, D., Calero, C., Femmer, H., Franch, X.: Systematic Mapping Study on Software Engineering for Sustainability (SE4S). In: Proceedings of the 18th International Conference on Evaluation and Assessment in Software Engineering. p. 14:1–14:14. ACM, New York, NY, USA (2014).
8. Penzenstadler, B., Bauer, V., Calero, C., Franch, X.: Sustainability in software engineering: A systematic literature review. In: 16th International Conference on Evaluation Assessment in Software Engineering (EASE 2012). pp. 32–41 (2012).
9. Petersen, K., Vakkalanka, S., Kuzniarz, L.: Guidelines for conducting systematic mapping studies in software engineering: An update. Inf. Softw. Technol. 64, 1–18 (2015).
10. Dick, M., Drangmeister, J., Kern, E., Naumann, S.: Green software engineering with agile methods. In: 2013 2nd International Workshop on Green and Sustainable Software (GREENS). pp. 78–85 (2013).
11. Chen, F., Schneider, J.-G., Yang, Y., Grundy, J., He, Q.: An Energy Consumption Model and Analysis Tool for Cloud Computing Environments. In: Proceedings of the First International Workshop on Green and Sustainable Software. pp. 45–50. IEEE Press, Piscataway, NJ, USA (2012).
12. Ardito, L., Morisio, M.: Green IT – Available data and guidelines for reducing energy consumption in IT systems. Sustain. Comput. Inform. Syst. 4, 24–32 (2014).
13. Curry, E., Donnellan, B.: Sustainable Information Systems and Green Metrics. In: Murugesan, S. and Gangadharan, G.R. (eds.) Harnessing Green It. pp. 167–198. John Wiley & Sons, Ltd, Chichester, UK (2012).
14. Amri, R., Saoud, N.B.B.: Towards a Generic Sustainable Software Model. In: 2014 Fourth International Conference on Advances in Computing and Communications (ICACC). pp. 231–234 (2014).

15. Doerflinger, J., Dearden, A., Gross, T.: A Software Development Methodology for Sustainable ICTD Solutions. In: CHI '13 Extended Abstracts on Human Factors in Computing Systems. pp. 2371–2374. ACM, New York, NY, USA (2013).
16. Murugesan, S., Gangadharan, G.R.: Green IT: An Overview. In: Murugesan, S. and Gangadharan, G.R. (eds.) Harnessing Green It. pp. 1–21. John Wiley & Sons, Ltd, Chichester, UK (2012).
17. Mahmoud, S.S., Ahmad, I.: A green model for sustainable software engineering. Int. J. Softw. Eng. Its Appl. 7, 55–74 (2013).
18. Noureddine, A., Bourdon, A., Rouvoy, R., Seinturier, L.: A preliminary study of the impact of software engineering on GreenIT. In: 2012 First International Workshop on Green and Sustainable Software (GREENS). pp. 21–27 (2012).
19. Zalewski, J., Sybramanian, N.: Developing a Green Computer Science Program. In: 2015 Seventh Annual IEEE Green Technologies Conference (GreenTech). pp. 95–102 (2015).
20. Ferri, J., Barros, R.M. d, Brancher, J.D.: Proposal for a Framework Focus on Sustainability. In: Chilean Computer Science Society (SCCC), 2011 30th International Conference of the. pp. 127–134 (2011).
21. Gu, Q., Lago, P., Muccini, H., Potenza, S.: A Categorization of Green Practices Used by Dutch Data Centers. Procedia Comput. Sci. 19, 770–776 (2013).
22. Kern, E., Dick, M., Naumann, S., Hiller, T.: Impacts of software and its engineering on the carbon footprint of ICT. Environ. Impact Assess. Rev. 52, 53–61 (2015).
23. Stefan, N., Eva, K., Markus, D.: Classifying Green Software Engineering - The GREENSOFT Model. (2013).
24. Scheller, R.M., Sturtevant, B.R., Gustafson, E.J., Ward, B.C., Mladenoff, D.J.: Increasing the reliability of ecological models using modern software engineering techniques. Front. Ecol. Environ. 8, 253–260 (2010).
25. Svetinovic, D.: Strategic requirements engineering for complex sustainable systems. Syst. Eng. 16, 165–174 (2013).
26. Albertao, F.: Sustainable Software Development. In: Murugesan, S. and Gangadharan, G.R. (eds.) Harnessing Green It. pp. 63–83. John Wiley & Sons, Ltd, Chichester, UK (2012).
27. Penzenstadler, B.: Towards a Definition of Sustainability in and for Software Engineering. In: Proceedings of the 28th Annual ACM Symposium on Applied Computing. pp. 1183–1185. ACM, New York, NY, USA (2013).
28. Koziolek, H.: Sustainability Evaluation of Software Architectures: A Systematic Review. In: Proceedings of the Joint ACM SIGSOFT Conference – QoSA and ACM SIGSOFT Symposium – ISARCS on Quality of Software Architectures – QoSA and Architecting Critical Systems – ISARCS. pp. 3–12. ACM, New York, NY, USA (2011).
29. Lago, P., Kazman, R., Meyer, N., Morisio, M., Müller, H.A., Paulisch, F.: Exploring Initial Challenges for Green Software Engineering: Summary of the First GREENS Workshop, at ICSE 2012. SIGSOFT Softw Eng Notes. 38, 31–33 (2013).
30. Burger, P., Christen, M.: Towards a capability approach of sustainability. J. Clean. Prod. 19, 787–795 (2011).
31. Penzenstadler, B., Fleischmann, A.: Teach sustainability in software engineering? In: 2011 24th IEEE-CS Conference on Software Engineering Education and Training (CSEE T). pp. 454–458 (2011).
32. Ardito, L., Procaccianti, G., Torchiano, M., Vetrò, A.: Understanding Green Software Development: A Conceptual Framework. IT Prof. 17, 44–50 (2015).

Organizational Maturity Models Architectures: A Systematic Literature Review

Viviana Saavedra[1], Abraham Dávila[2], Karin Melendez[2], and Marcelo Pessoa[3]

[1]Escuela de Post-grado, Pontificia Universidad Católica del Perú, Lima, Perú
v.saavedra@pucp.edu.pe
[2]Departmento de Ingeniería, Pontificia Universidad Católica del Perú, Lima, Perú
{abraham.davila, kmelendez}@pucp.edu.pe
[3]Polytechnic School, University of Sao Paulo, Sao Paulo, Brazil
mpessoa@usp.br

Abstract. Maturity models (MMs) are tools to assess how reliable is an organization and to identify its strengths and weaknesses. The increasing number of MMs developed for several domains or specific contexts based on different architectural styles (AS's), makes more difficult its understanding and reuse. This paper aims to identify and compare the AS's of organizational MMs for different domains. A Systematic Literature Review (SLR) was conducted. The SLR included 70 studies that describe the architecture of MMs for different contexts. A classification scheme was defined and tested with a group of MMs. Another group of MMs adopted the AS of existing models. As a result, the MMs found were classified into nine AS's. The AS's derived from SW-CMM and CMMI-staged are quite similar. The ISO-SPICE-based and the OPM3-based architectures are the most different. The AS's most used in studies were the CMMI-based and the progression-staged model.

Keywords: Maturity Model, Organizational Maturity, Maturity Model Architecture, Maturity Model Development.

1 Introduction

The concept of maturity model (MM) was first applied in the late 1970s, in the Crosby's maturity grid for quality management [1]. Later, this grid along with the quality principles proposed in the 1930s served as the foundation for the Capability Maturity Model for Software (SW-CMM) published in the 1990s by the Software Engineering Institute (SEI), which collected best practices to improve software processes [2]. Since then, many MMs were created based on the SW-CMM model [1], like the Systems Engineering Capability Maturity Model (SE-CMM) [3], the Test Maturity Model Integration (TMMi) [4], and the Capability Maturity Model Integration (CMMI) [5]. Due to the SW-CMM success, the MMs began to be developed or adapted for different contexts in the industry as a method for continuous improvement, self-assessment or benchmarking [6]. Such is the case of the P-CMM model [1], the

© Springer International Publishing AG 2017
J. Mejia et al. (eds.), *Trends and Applications in Software Engineering*, Advances in Intelligent Systems and Computing 537, DOI 10.1007/978-3-319-48523-2_4

OPM3 model [7], the BPMMM model [8], among others. This proliferation of MMs and the rise of publications about the subject are generating confusion about the characteristics of each proposal, making difficult their understanding and reuse [6]. This paper provides the answer to a first group of research questions of a Systematic Literature Review (SLR) which aims to identify and compare the architectural styles (AS's) of organizational MMs for different domains.

The remainder of the paper is organized as follows: Section 2 presents the background on the subject addressed and related work; Section 3 describes the research method; Section 4 presents the analysis of the findings; and Section 5 presents the final discussion and proposals for future works.

2   Background and Related Work

Maturity models (MMs) are defined as a set of elements arranged in an evolutionary path with measurable transitions between levels [9]. In general, MMs are composed by the following elements [10]: (i) a number of maturity levels, typically between 3 and 6; (ii) a name for each level, such as CMM levels: initial, repeatable, defined, managed and optimizing [2]; (iii) A general description of each level; (iv) a number of dimensions to be assessed, such as the CMM key process areas [2]; (v) a number of elements or activities for each dimension, such as the goals and key practices for each CMM key process area [2]; and (vi) a description of each element or activity.

Van Steenbergen et al. in 2008 [11] classify existing MMs into three basic types: (i) Staged fixed-level models, which have a fixed number of maturity levels, and a specific number of focus areas defined for each level; (ii) Continuous fixed-level models, which have also a fixed number of maturity levels and a specific number of focus areas, but all the maturity levels are distinguished on each focus area; and (iii) Focus Area Oriented models, where each focus area has its own specific maturity levels. Another classification scheme was presented in 2012 by Caralli [9], where propose three general types of MMs: (i) Progression models, which represent the simple evolution of the assessed element; (ii) Capability Maturity Models, where the transition between the maturity levels indicates the evolution of the capability of the organization in relation to the subject matter of the MM; and (iii) Hybrid models, which represent the evolution of an element like in the progression model but the transition between states has the institutionalizing features of the capability MM.

In the software engineering (SE) domain, the term architecture is defined as the essential organization of a system, represented by its elements, the relationship between them and their context, and the rules that guide its design and evolution [12]. When systems share a similar set of topics, this similarity is described as an architectural style (AS) [13]. It is a good practice to reuse an AS to take advantage of its previous experience [13]. In the context of MMs, knowing their AS's may serve as a reference to choose reusing validated MMs instead of developing new ones.

On a preliminary non-systematic review and during this SLR the following related studies were found. Becker et al. [14] published in 2009 an analysis of 51 MMs of which six were chosen as a reference to propose a model to design MMs for IT management. The study of von Wangenheim et al. [15] presented in 2010 a systematic review where 52 capability/maturity models related to software processes were identified; it was found that most of them were developed based on SW-CMM, ISO/IEC 15504 (SPICE) and CMMI-DEV. The systematic review of García-Mireles [16], presented in 2012, analyze the existing methods and recommended practices for MMs development. Helgesson et al. [17] published in 2012 a summary of the methods found in the literature to evaluate MMs for process improvement. The mapping study of Wendler [18] presented an overview about MMs research topics, application domains and development and validation processes in studies until 2010. Kohlegger et al. [19] published in 2009 a comparison of 16 MMs using structured content analysis, and proposed a verification list to be used in the development of MMs. The study of Khoshgoftar and Osman [20] presented in 2009 a comparison of project management MMs based on 27 general variables common among all models. The previous studies mainly focus on the analysis of methods for development, evaluation or validation of MMs, most of them for a specific domain, unlike this SLR which focuses on identifying and comparing the AS's of MMs for different domains, with the purpose of finding evidence about the coverage of MMs from SE to other domains. This SLR includes recent studies published until 2015, which are not covered in the previous work.

3   Research Method

This section describes the activities undertaken to conduct this SLR, following as a basis the guidelines of Kitchenham and Chartes [21].

3.1   Research Purpose and Questions

This study aims to identify the AS's of organizational MMs for different domains, and their relevant characteristics that allow comparing them. Table 1 presents the research questions (RQ). The primary RQ intends to compare the AS's identified, and it is decomposed in two secondary questions to identify the characteristics of the AS's.

Table 1. Research questions

Research Questions	Motivation
RQ1. What are the main similarities and differences between the architectural styles identified?	To compare AS's identified to find similarities and differences between them.
RQ1.1. What elements compose the maturity models described in the studies?	To identify the elements composing the architecture of MM's.
RQ1.2. What style of architecture have the maturity models described in the studies?	To identify the main characteristics of the AS's used to define maturity models.

3.2   Review Protocol

The review protocol is a formal specification of the methods that will be used to conduct a systematic review [21]. This section describes the review protocol defined for this study.

Search Strategy. It was used the approach proposed by Santos et al. [22] to define the search terms. It consists in identifying keywords derived from the research questions for each of the following elements: population (P), intervention (I), comparison (C) and outcomes (O). The terms selected are linked according to the structure P AND I AND C AND O. In this study, the identified keywords "maturity model", "architecture" and "description", and their related terms and synonyms were considered to define the search terms (see Table 2, where the column "E" is the element of the PICO strategy). For some terms related to the keyword "maturity model" like "ISO/IEC 15504-7", "CMM for Software" and "CMMI for Development" was considered the principal term only (15504-7, CMM, and CMMI) to prevent that relevant articles were discarded.

Four relevant studies were identified manually [23–26] and were used to test the search terms. After an iterative process, the final search terms were defined. The search string was adapted to each digital library. The final search took place between September and October 2015. The following digital libraries were selected due to their relevance in research in the fields of SE [27] and information systems (IS) [28]: IEEE Xplore, ScienceDirect, Scopus, Web of Science, Wiley Online Library, EBSCOhost Web,Emerald Insight and ProQuest. Grey literature from the SEI digital library was also included because of its significance in the subject of this SLR as the creator of SW-CMM and CMMI models (http://resources.sei.cmu.edu/library/).

Table 2. Search terms

E	Description	Related Terms
P	Set of entities that will be reviewed. In this study this element is represented by maturity models.	"maturity model" OR CMM OR CMMI OR "15504-7" OR "organizational maturity" OR "organizational profile" OR "maturity profile"
I	What will be evaluated in the population. In this study this element is represented by architecture of maturity models.	architecture OR structure OR taxonomy OR configuration OR constitution
C	Elements that will be a pattern or reference for comparison of the results.	Not applied in this study since the results will not be compared to any pattern.
O	Information expected to be found in the search. This study expects to find documents with a description of the architecture of maturity models.	description OR characterization OR documentation OR specification OR scheme OR model OR pattern OR map

Selection Criteria and Quality Assessment. The inclusion and exclusion criteria are shown in Table 3. Date limitation was not considered since this SLR tried to cover as many papers as possible. To assess the quality of the primary studies it was defined the questionnaire shown in Table 4, adapted from the studies of Williams and Carver [29] and Rouhani et al. [30]. The questions were scored with a rating scale taken from the study of Rouhani et al. [30]: Yes (Y) = 1, Partly (P) = 0.5 or No (N) = 0.

Table 3. Inclusion and exclusion criteria

Item	Criteria	Type
IC1	Studies that focused on describing the architecture of a maturity model for organizations in any domain	Inclusion
EC1	Studies whose title was not related to the subject of the SLR.	Exclusion
EC2	Studies that having relevant words in the title, the abstract or the content was not related to the subject of the SLR.	Exclusion
EC3	Studies written in other languages than English or Spanish.	Exclusion
EC4	Articles related to maturity models but not including the architecture specification of any model.	Exclusion
EC5	Duplicated items.	Exclusion
EC6	Successive articles covering the same subject of study; the most complete version of the item was included.	Exclusion
EC7	Books, secondary studies, tertiary studies, summary of conference proceedings and newspaper articles.	Exclusion

Table 4. Quality assesment questionnaire

No.	Question
1	Is the MM in the study defined in detail?
2	Is the domain or context of the study clearly stated?
3	Are the limitations of the study described clearly?
4	Does the study mention contributions for the scientific, academic or industry communities?
5	Do the findings contribute to answer the research questions of this SLR?

Data Extraction. A form with the following data was filled per each article.

- Bibliographic information: source digital library, article type, article title, year of publication, author(s) and publication or venue name.
- MM information: name; elements conforming the model according to the definition of Fraser et al. [10] (see Section 2); type according to Caralli's classification [9] and type according to Van Steenbergen's classification [11] (see Section 2); domain; model or standard used as a reference; contribution of the study.

Threats to Validity. The threats to the validity of the study are described below.

- Data sources. The most popular digital libraries in research on SE and IS were selected for the searches. Therefore, this study may have not included some relevant studies that are not indexed in the sources considered.
- Studies selection. Some stages of the selection process may have been based on personal judgment, so there is a risk of a biased selection. However the selection criteria were defined to guide the process and reduce the bias.
- Quality assessment. It was found that the studies with less than 50% of quality score had enough information to answer the research questions. Therefore they were included despite their low quality scoring.
- Analysis of results. The classification scheme tried in this study was obtained combining two generic MMs typologies. Therefore, there is a risk that some AS may have been overlooked in the classification.

4   Analysis of Results

This section presents the result of the SLR described in Section 3. To obtain the primary studies it was performed the four stages selection process described in Fig. 1. In the fourth stage, some articles were excluded because the information related to the MM architecture was reduced and did not contribute to answer the research questions. The initial search produced a result of 2540 studies; after the four stages, 70 primary studies were obtained. The list of selected articles is given in Appendix A. Table 5 presents the score obtained after the quality assessment. More than 90% of studies (65 of 70) had a score equal to or greater than 50% of the maximum score.

Table 5. Quality assesment of selected studies

Score	No	Studies	Score	No	Studies
4.5	6	[S18,S24,S45,S47,S55,S60]	3.0	17	[S01,S02,S08,S11,S26,S29,S30-S33,S38,S40,S46,S49, S54,S62,S66]
4.0	16	[S05,S06,S17,S22,S25,S28,S39, S41,S42,S44,S56-S59, S68,S69]	2.5	8	[S13,S16,S19-S21,S48, S64,S65]
3.5	18	[S03,S04,S07,S09,S12,S14,S15, S23,S27,S36,S37,S43, S50-S53,S61,S67]	2.0	5	[S10,S34,S35,S63,S70]

The domains with more than one studies were: Construction industry [S04,S06,S19,S51]; Cybersecurity [S10,S68]; Education [S13,S22]; IT related [S07,S11,S20,S23, S25,S26,S38,S40,S57,S66]; IT Services [S18,S33,S44, S58,S61]; IT/SW Products and services acquisition [S02,S56]; Knowledge management [S01,S08,S31, S43,S63]; Logistics [S52,S67]; Software engineering related [S09,S12,S14,S21,S24,S36,S39,S41,S47-S50,S54,S55,S64, S65]; Software product line engineering [S28,S46].

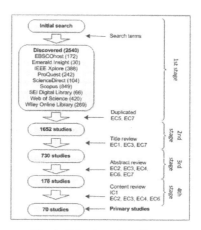

Fig. 1. Selection of studies

4.1   RQ1.1 What elements compose the maturity models described in the studies?

The elements described in Section 2 were identified in the selected studies as a part of the MM, the results were the following:

- Maturity levels: The 100% of selected studies presented this element. Values vary between 2 and 6, but 5 maturity levels is the most common value (39 studies).
- Name of maturity levels: In 86% of selected studies was defined a name for maturity levels. Most of the MMs (46 of 70) created their own denomination for the levels, but some models take the name from other better known models [6, 17] like CMMI [S14,S30,S36,S61], CMM [S02,S16,S25,S43] and ISO-SPICE [S49].
- Key dimensions to be assessed: All selected studies presented this element. The most frequently mentioned key dimensions are Key Process Areas (24 studies) and Key Processes (12 studies). Other key dimensions less frequent are Capabilities, Goals, Practices, Activities, Attributes, Critical factors, among others.
- Elements or activities for each dimension: In 81% of selected studies it was found this element. Most of the studies define for this element: Key practices, Best practices, Specific practices Base practices, Generic practices, Goals and Key indicators.

4.2   RQ1.2. What style of architecture have the maturity models described in the studies?

Analyzing the MMs described in the selected studies, two trends were noticed. In 30 of 70 studies it was adopted an AS similar to four better known models [6, 17]: SW-CMM, CMMI, ISO SPICE and OPM3. The MMs in these studies had the same AS than the original reference model with small changes to fit their specific context. The studies of this group were classified into four AS's, named "An", which are listed in Table 6. Despite their similarities, SW-CMM and CMMI were classified separately because of the two evolution representations available in CMMI.

In a second group of studies (40 of 70) the AS of the MMs were based on the specific requirements of the domain and some characteristics from the models or standards of reference. Due to the variety found in these studies, it was chosen to define a generic classification scheme combining the typologies of Caralli [9] and Van Steenbergen et al. [11] described in Section 2, and test it classifying the AS's of this group. Given that both proposals are generic, they were found suitable to classify MMs from different domains. With the scheme defined, the information extracted from these studies allow to classify them into five generic AS's, named "GAn", which are listed in Table 6.

4.3 RQ1. What are the main similarities and differences between the architectural styles identified?

A brief description of the AS's identified is shown in Table 7. The main similarities and differences are shown in Table 8. Appendix B[1] shows a comparison chart of the AS's identified.

Table 6. Studies per architectural style (AS)

Id	Description	No	Studies
A1	SW-CMM-based model	8	[S04,S09,S16,S31,S34,S51,S55,S59]
A2	CMMI-based model	17	Staged/Continuous: [S23,S56-S58] Staged:[S07,S22,S30,S32,S36,S38,S40, S53] Continuous: [S10,S14,S49,S50,S61]
A3	ISO-SPICE-based model	04	[S42,S45,S69,S70]
A4	OPM3-based model	01	[S19]
GA1	Progression-staged model	16	[S01,S03,S05,S12,S13,S17,S18,S25, S29,S35,S39,S41,S48,S62-S64]
GA2	Progression-continuous model	11	[S15,S21,S27,S28,S33,S37,S46,S52, S54,S60,S65]
GA3	Progression-focus area oriented model	05	[S11,S20,S44,S47,S66]
GA4	Capability maturity-staged model	05	[S08,S24,S26,S67,S68]
GA5	Capability maturity-continuous model	03	[S02,S06,S43]

5   Final Discussion and Future Work

The largest number of publications in selected studies was observed between 2009 and 2010, coinciding with the publication date of the model CMMI-DEV (2010). The domains that have the largest number of studies are SE and information technology (IT). But it was also found MMs for other contexts, like knowledge management, construction, education, medical, and logistics, among others. The SW-CMM and CMMI models are the most frequent references mentioned in the studies, coinciding with the result of the study of von Wangenheim et al. [15]; indicating they are important references for MMs development [15].

Nine AS's were identified in selected studies. Four AS's derived from the architecture of other MMs (SW-CMM, CMMI, ISO SPICE and OPM3) with minimal changes. This indicates that existing MMs can be adapted to other contexts in the industry, like construction [S04, S19, and S51], environmental management [S16], education [S22] or ship design [S30]. Five AS's were identified when trying a classification scheme defined by combining two existing MMs typologies.

The findings showed that most AS's have continuous representation, indicating that most studies aim to assess the evolution of focus areas individually. The AS's derived from SW-CMM and CMMI-staged are very similar; and the AS derived from ISO-SPICE and OPM3 are the most different. The AS derived from CMMI was the most used in studies. It is possible that being CMMI the integration of several MMs, it is more adaptable to a variety of contexts [5]. However, there is a risk that the search terms "CMM" and "CMMI" may have influenced this result. The second most used AS was Progression-staged model, indicating that many MMs represent a simple evolution of the subject of interest, through a staged scale.

---

[1] Available in https://drive.google.com/drive/folders/0B_K_llz5juqPWWI3bEFaUlVPTmM

Table 7. Brief description of AS's identified

AS	Description
A1 (SW-CMM-based model)	Evolution of the capability of the key processes areas throughout the organization. Similar to SW-CMM.s.
A2 (CMMI-based model)	Evolution of the capability of the processes areas throughout the organization. Similar to CMMI
A3 (ISO-SPICE-based model)	Evolution of an individual process. Similar to ISO-SPICE.
A4 (OPM3-based model)	Evolution of the processes improvement throughout all the domains. Similar to OPM3.
GA1 (Progression-staged model)	Simple evolution of the maturing element. To achieve a maturity level, all focus areas defined for that level must be implemented successfully.
GA2 (Progression-continuous model)	Simple evolution of the maturing element. The maturity is assessed for all focus areas individually through the same number of levels.
GA3 (Progression-focus area oriented model)	Simple evolution of the maturing element. The maturity is assessed for all focus areas individually with their own levels scale.
GA4 (Capability maturity-staged model)	Evolution of the capability of the assessed element throughout the organization. To achieve a maturity level, all focus areas defined for that level must be implemented successfully
GA5 (Capability maturity-continuous model)	Evolution of the capability of the assessed element throughout the organization. The maturity is assessed for all focus areas individually through the same number of levels.

Table 8. Main similarities and differences between the AS's

Similarities	Differences
A1, A2, A3, A4, GA1, GA2, and GA4 have a fixed number of maturity levels.	GA3 defines a specific level scale for each focus area (focus area oriented); being used when each focus area of the MM progress differently.
A1, A2, GA1, GA2, GA3 and GA4 define one dimension to be evaluated.	
A1 and A2-Staged are quite similar in their elements and their evolution representation; therefore, it can be considered that A1 is covered by A2.	A2 is the only AS with staged and continuous representations in the same model.
A1, A2, GA1 and GA4 have staged representation, meaning that to achieve a maturity level all focus areas defined for that level must be implemented successfully.	A3 and A4 define more than one dimension, 2 and 3 respectively. Both present the domain as a separated dimension, which facilitates their adaptation to different do-mains.
A2, A3, A4, GA2, GA3 and GA5 have continuous representation, meaning that the maturity is assessed for all focus areas individually through all levels.	
For A1, A2, GA4 and GA5, maturity represents the evolution of the capability of the assessed element throughout the organization.	Despite GA4 has similar evolution representation than A1 and A2-staged, it present different elements and number of levels.
For A3, A4, GA1, GA2 and GA3, maturity represents the simple evolution of the maturing element.	

To enhance the SLR findings, more research has to be done to find out whether the MMs identified have been validated in the domains they were developed for. Also to identify other AS's and patterns different to those covered in this study. The research can be extended to include models from other contexts in the industry used for categorizing organizations or benchmarking, such as the case of the hotels classification systems, for instance.

Acknowledgements. This work is framed within the ProCal-ProSer project funded by Innóvate Perú under contract 210-FINCYT-IA-2013 and partially supported by the Department of Engineering and the Grupo de Investigación y Desarrollo de Ingeniería de Software (GIDIS) from the Pontificia Universidad Católica del Perú.

References

1. Paulk, M.C.: A history of the capability maturity model for software. ASQ Softw. Qual. Prof. 12, 5–19 (2009).
2. Paulk, M., Curtis, W., Chrissis, M.B., Weber, C.: Capability Maturity Model for Software (Version 1.1). Software Engineering Institute, Carnegie Mellon University, Pittsburgh, PA (1993).
3. Bate, R., Garcia, S., Armitage, J., Cusick, K., Jones, R., Kuhn, D., Minnich, I., Pierson, H., Powell, T., Reichner, A.: A Systems Engineering Capability Maturity Model, Version 1. Software Engineering Institute, Carnegie Mellon University, Pittsburgh, Pennsylvania (1994).
4. van Veenendaal, E., Cannegieter, J.J.: Test Maturity Model integration (TMMi). TMMi Found. (2010).
5. CMMI Product Team: Capability Maturity Model® Integration (CMMI SM), Version 1.1. CMMI Syst. Eng. Softw. Eng. Integr. Prod. Process Dev. Supplier Sourc. CMMI-SESWIPPDSS V1 1. (2002).
6. Mettler, T., Rohner, P., Winter, R.: Towards a classification of maturity models in information systems. In: Management of the Interconnected World - ItAIS: The Italian Association for Information Systems. pp. 333–340 (2010).
7. Cooke-Davies, T.J., Arzymanow, A.: The maturity of project management in different industries: An investigation into variations between project management models. Int. J. Proj. Manag. 21, 471 – 478 (2003).
8. Rosemann, M., De Bruin, T.: Towards a Business Process Management Maturity Model. In: ECIS 2005 Proceedings of the Thirteenth European Conference on Information Systems. pp. 1–12. Verlag and the London School of Economics, Germany, Regensburg (2005).
9. Caralli, R.A.: Discerning the Intent of Maturity Models from Characterizations of Security Posture. Software Engineering Institute, Carnegie Mellon University, Pittsburgh, PA (2012).
10. Fraser, P., Moultrie, J., Gregory, M.: The use of maturity models/grids as a tool in assessing product development capability. In: Engineering Management Conference, 2002. IEMC '02. 2002 IEEE International. pp. 244–249 vol.1 (2002).
11. Van Steenbergen, M.., Van Den Berg, M.., Brinkkemper, S..: A balanced approach to developing the enterprise architecture practice. Lect. Notes Bus. Inf. Process. 12 LNBIP, 240–253 (2008).
12. IEEE Recommended Practice for Architectural Description of Software-Intensive Systems. IEEE Std 1471-2000. i–23 (2000).
13. Eeles,   P.:   What   is   a   software   architecture?, http://www.ibm.com/developerworks/rational/library/feb06/eeles/.
14. Becker, J., Knackstedt, R., Pöppelbuß, J.: Developing Maturity Models for IT Management. Bus. Inf. Syst. Eng. 1, 213–222 (2009).
15. von Wangenheim, C.G., Hauck, J.C.R., Salviano, C.F., von Wangenheim, A.: Systematic literature review of software process capability/maturity models. In: Proceedings of International Conference on Software Process Improvement and Capability Determination (SPICE), Pisa, Italy (2010).

16. García-Mireles, G.A., Moraga, M.Á., García, F.: Development of maturity models: a systematic literature review. IET Conf. Proc. 279–283(4) (2012).

17. Helgesson, Y.Y.L., Höst, M., Weyns, K.: A review of methods for evaluation of maturity models for process improvement. J. Softw. Evol. Process. 24, 436–454 (2012).

18. Wendler, R.: The maturity of maturity model research: A systematic mapping study. Inf. Softw. Technol. 54, 1317–1339 (2012).

19. Kohlegger, M., Maier, R., Thalmann, S.: Understanding Maturity Models. Results of a Structured Content Analysis. In: Proceedings of I-KNOW 2009 International Conference on Knowledge Management and Knowledge Technologies. pp. 51–61. , Austria (2009).

20. Khoshgoftar, M., Osman, O.: Comparison of maturity models. In: Proceedings - 2009 2nd IEEE International Conference on Computer Science and Information Technology, ICCSIT 2009. pp. 297–301 (2009).

21. Kitchenham, B., Charters, S.: Guidelines for performing Systematic Literature Reviews in Software Engineering. EBSE Technical Report EBSE-2007-01, Staffordshire (2007).

22. Santos, C.M. da C., Pimenta, C.A. de M., Nobre, M.R.C.: A estratégia PICO para a construção da pergunta de pesquisa e busca de evidências. Rev. Lat. Am. Enfermagem. 15, 508 – 511 (2007).

23. Burnstein, I., Suwanassart, T., Carlson, R.: Developing a Testing Maturity Model for software test process evaluation and improvement. In: Test Conference, 1996. Proceedings., International. pp. 581–589 (1996).

24. Fontana, R.M.. b, Meyer, J.., V., Reinehr, S.., Malucelli, A..: Progressive Outcomes: A framework for maturing in agile software development. J. Syst. Softw. 102, 88–108 (2015).

25. Mettler, T., Blondiau, A.: HCMM - A maturity model for measuring and assessing the quality of cooperation between and within hospitals. In: Proceedings - IEEE Symposium on Computer-Based Medical Systems (2012).

26. Wetering, R. van de, Batenburg, R.: A {PACS} maturity model: A systematic meta-analytic review on maturation and evolvability of {PACS} in the hospital enterprise. Int. J. Med. Inf. 78, 127 – 140 (2009).

27. Zhang, H., Babar, M.A., Tell, P.: Identifying relevant studies in software engineering. Inf. Softw. Technol. 53, 625 – 637 (2011).

28. Neff, A.A., Hamel, F., Herz, T.P., Uebernickel, F., Brenner, W., Brocke, J. vom: Developing a maturity model for service systems in heavy equipment manufacturing enterprises. Inf. Manage. 51, 895 – 911 (2014).

29. Williams, B.J., Carver, J.C.: Characterizing software architecture changes: A systematic review. Inf. Softw. Technol. 52, 31 – 51 (2010).

30. Rouhani, B.D., Mahrin, M.N., Nikpay, F., Ahmad, R.B., Nikfard, P.: A systematic literature review on Enterprise Architecture Implementation Methodologies. Inf. Softw. Technol. 62, 1 – 20 (2015).

Appendix A. List of Selected Studies

S01. Kaner, M., Karni, R.: A Capability Maturity Model for Knowledge-Based Decisionmaking. Inf. Knowl. Syst. Manag. 4, 225 – 252 (2004).

S02. Baker, E.R., Cooper, L.: Software acquisition management maturity model (SAM...). Program Manag. 23, 43 (1994).

S03. Alvarez, R.L.P., Martin, M.R., Silva, M.T.: Applying the maturity model concept to the servitization process of consumer durables companies in Brazil. J. Manuf. Technol. Manag. 26, null (2015).

S04. Sarshar, M., Haigh, R., Amaratunga, D.: Improving project processes: best practice case study. Constr. Innov. 4, 69–82 (2004).

S05. Masalskyte, R., Andelin, M., Sarasoja, A.-L., Ventovuori, T.: Modelling sustainability maturity in corporate real estate management. J. Corp. Real Estate. 16, 126–139 (2014).

S06. Willis, C.J., Rankin, J.H.: The construction industry macro maturity model (CIM3): theoretical underpinnings. Int. J. Product. Perform. Manag. 61, 382–402 (2012).

S07. Ibrahim, L., Pyster, A.: A single model for process improvement. IT Prof. 6, 43–49 (2004).

S08. Harigopal, U., Satyadas, A.: Cognizant enterprise maturity model (CEMM). Syst. Man Cybern. Part C Appl. Rev. IEEE Trans. On. 31, 449–459 (2001).

S09. Burnstein, I., Suwanassart, T., Carlson, R.: Developing a Testing Maturity Model for software test process evaluation and improvement. In: Test Conference, 1996. Proceedings., International. pp. 581–589 (1996).

S10. Curtis, P.D., Mehravari, N.: Evaluating and improving cybersecurity capabilities of the energy critical infrastructure. In: Technologies for Homeland Security (HST), 2015 IEEE International Symposium on. pp. 1–6 (2015).

S11. Smits, D., Van Hillegersberg, J.: IT Governance Maturity: Developing a Maturity Model Using the Delphi Method. In: System Sciences (HICSS), 2015 48th Hawaii International Conference on. pp. 4534–4543 (2015).

S12. Tripathi, A.K., Ratneshwer: Some Observations on a Maturity Model for CBSE. In: Engineering of Complex Computer Systems, 2009 14th IEEE International Conference on. pp. 273–281 (2009).

S13. Manjula, R., Vaideeswaran, J.: A Bootstrap Approach of Benchmarking Organizational Maturity Model of Software Product With Educational Maturity Model. Int. J. Mod. Educ. Comput. Sci. 4, 50–58 (2012).

S14. Solemon, B., Sahibuddin, S., Ghani, A.A.A.: A New Maturity Model for Requirements Engineering Process: An Overview. J. Softw. Eng. Appl. 5, 340–350 (2012).

S15. Hillson, D.: Assessing organisational project management capability. J. Facil. Manag. 2, 298–311 (2003).

S16. Doss, D.A., Kamery, R.H.: The Capability Maturity Model: A valid architecture to support a baseline environmental management maturity model. Allied Acad. Int. Conf. Acad. Inf. Manag. Sci. Proc. 9, 1–5 (2005).

S17. Wetering, R. van de, Batenburg, R.: A {PACS} maturity model: A systematic meta-analytic review on maturation and evolvability of {PACS} in the hospital enterprise. Int. J. Med. Inf. 78, 127 – 140 (2009).

S18. Neff, A.A., Hamel, F., Herz, T.P., Uebernickel, F., Brenner, W., Brocke, J. vom: Developing a maturity model for service systems in heavy equipment manufacturing enterprises. Inf. Manage. 51, 895 – 911 (2014).

S19. Jia, G., Chen, Y., Xue, X., Chen, J., Cao, J., Tang, K.: Program management organization maturity integrated model for mega construction programs in China. Int. J. Proj. Manag. 29, 834 – 845 (2011).

S20. Van Steenbergen, M.., Van Den Berg, M.., Brinkkemper, S..: A balanced approach to developing the enterprise architecture practice. Lect. Notes Bus. Inf. Process. 12 LNBIP, 240–253 (2008).

S21. Kang, S.., Myung, J.., Yeon, J.., Ha, S.-W.., Cho, T.., Chung, J.-M.., Lee, S.-G..: A general maturity model and reference architecture for SaaS service. Lect. Notes Comput. Sci. Subser. Lect. Notes Artif. Intell. Lect. Notes Bioinforma. 5982 LNCS, 337–346 (2010).

S22. Solar, M.., Sabattin, J.., Parada, V..: A maturity model for assessing the use of ICT in school education. Educ. Technol. Soc. 16, 206–218 (2013).

S23. De Sousa Pereira, R.F., Da Silva, M.M.: A maturity model for implementing ITIL v3. In: Proceedings - 2010 6th World Congress on Services, Services-1 2010. pp. 399–406 (2010).

S24. Niazi, M., Wilson, D., Zowghi, D.: A maturity model for the implementation of software process improvement: An empirical study. J. Syst. Softw. 74, 155–172 (2005).

S25. Tan, C.-S., Sim, Y.-W., Yeoh, W.: A maturity model of enterprise business intelligence. In: Knowledge Management and Innovation: A Business Competitive Edge Perspective - Proceedings of the 15th International Business Information Management Association Conference, IBIMA 2010. pp. 20–29 (2010).

S26. Solar, M.., Concha, G.., Meijueiro, L..: A model to assess open government data in public agencies. Lect. Notes Comput. Sci. Subser. Lect. Notes Artif. Intell. Lect. Notes Bioinforma. 7443 LNCS, 210–221 (2012).

S27. Mayer, J., Fagundes, L.L.: A model to assess the maturity level of the risk management process in information security. In: 2009 IFIP/IEEE International Symposium on Integrated Network Management-Workshops, IM 2009. pp. 61–70 (2009).

S28. Ahmed, F., Capretz, L.F.: An architecture process maturity model of software product line engineering. Innov. Syst. Softw. Eng. 7, 191–207 (2011).

S29. Sukhoo, A., Barnard, A., Eloff, M.M., Van Der Poll, J.A.: An Evolutionary Software Project Management Maturity Model for Mauritius. Interdiscip. J. Inf. Knowl. Manag. 2, 99–118 (2007).

S30. Caracchi, S.., Sriram, P.K.., Semini, M.., Strandhagen, J.O..: Capability Maturity Model Integrated for Ship Design and Construction. IFIP Adv. Inf. Commun. Technol. 440, 296–303 (2014).

S31. Huffman, J.., Whitman, L.E..: Developing a capability maturity model for enterprise intelligence. In: IFAC Proceedings Volumes (IFAC-PapersOnline). pp. 13086–13091 (2011).

S32. Machado, C.G.., Pinheiro De Lima, E.. b, Gouvea Da Costa, S.E.. b, Cestari, J.M.A.P.., Kluska, R.A.., Hundzinski, L.N..: Developing a sustainable operations maturity model (SOMM). In: 22nd International Conference on Production Research, ICPR 2013 (2013).

S33. Wendler, R.: Development of the organizational agility maturity model. In: 2014 Federated Conference on Computer Science and Information Systems, FedCSIS 2014. pp. 1197–1206 (2014).

S34. Haukijärvi, I.: E-Learning Maturity Model – Process-oriented assessment and improvement of e-Learning in a Finnish University of Applied Sciences. IFIP Adv. Inf. Commun. Technol. 444, 76–93 (2014).

S35. Grim, T.: Foresight maturity model (FMM): Achieving best practices in the foresight field. J. Futur. Stud. 13, 69–80 (2009).

S36. Erkollar, A., Zimmermann, A.: Framework for capability and maturity evaluation of service-oriented enterprise architectures. In: IMETI 2010 - 3rd International Multi-Conference on Engineering and Technological Innovation, Proceedings. pp. 273–278 (2010).

S37. Mettler, T., Blondiau, A.: HCMM - A maturity model for measuring and assessing the quality of cooperation between and within hospitals. In: Proceedings - IEEE Symposium on Computer-Based Medical Systems (2012).

S38. Santana Tapia, R.: ICoNOs MM: The IT-enabled collaborative networked organizations maturity model. IFIP Adv. Inf. Commun. Technol. 307, 591–599 (2009).

S39. Petrinja, E.., Nambakam, R.., Sillitti, A..: Introducing the opensource maturity model. In: Proceedings of the 2009 ICSE Workshop on Emerging Trends in Free/Libre/Open Source Software Research and Development, FLOSS 2009. pp. 37–41 (2009).

S40. Caballero, I.., Caro, A.., Calero, C.., Piattini, M..: IQM3: Information quality management maturity model. J. Univers. Comput. Sci. 14, 3658–3685 (2008).

S41. Rathfelder, C., Groenda, H.: ISOAMM: An independent SOA maturity model. Lect. Notes Comput. Sci. Subser. Lect. Notes Artif. Intell. Lect. Notes Bioinforma. 5053 LNCS, 1–15 (2008).

S42. Hefner, R.: Lessons learned with the systems security engineering capability maturity model. In: Proceedings - International Conference on Software Engineering. pp. 566–567 (1997).

S43. Jochem, R.. b, Geers, D.., Heinze, P..: Maturity measurement of knowledge-intensive business processes. TQM J. 23, 377–387 (2011).

S44. Rudolph, S., Krcmar, H.: Maturity model for IT service catalogues an approach to assess the quality of IT service documentation. In: 15th Americas Conference on Information Systems 2009, AMCIS 2009. pp. 6223–6232 (2009).

S45. Egan, I.. d, Ritchie, J.M.., Gardiner, P.D..: Measuring performance change in the mechanical design process arena. Proc. Inst. Mech. Eng. Part B J. Eng. Manuf. 219, 851–863 (2005).

S46. Lamas, E.., Ferreira, É.., Do Nascimento, M.R.., Dias, L.A.V.., Silveira, F.F..: Organizational testing management maturity model for a software product line. In: ITNG2010 - 7th International Conference on Information Technology: New Generations. pp. 1026–1031 (2010).

S47. Fontana, R.M.. b, Meyer, J.., V., Reinehr, S.., Malucelli, A..: Progressive Outcomes: A framework for maturing in agile software development. J. Syst. Softw. 102, 88–108 (2015).

S48. Kassou, M., Kjiri, L.: SOASMM: A novel service oriented architecture Security Maturity Model. In: Proceedings of 2012 International Conference on Multimedia Computing and Systems, ICMCS 2012. pp. 912–918 (2012).

S49. April, A.. d, Hayes, J.H.., Abran, A.., Dumke, R..: Software Maintenance Maturity Model (SMmm): The software maintenance process model. J. Softw. Maint. Evol. 17, 197–223 (2005).

S50. Patel, C., Ramachandran, M.: Story card Maturity Model (SMM): A process improvement framework for agile requirements engineering practices. J. Softw. 4, 422–435 (2009).

S51. Zhai, F., Liu, R.: Study on framework of construction project management maturity model. In: Proceedings - ICSSSM'07: 2007 International Conference on Service Systems and Service Management (2007).

S52. Battista, C.., Schiraldi, M.M..: The logistic maturity model: Application to a fashion company. Int. J. Eng. Bus. Manag. 5, (2013).

S53. Axelsson, J.: Towards a process maturity model for evolutionary architecting of embedded system product lines. In: ACM International Conference Proceeding Series. pp. 36–42 (2010).

S54. Niknam, M., Ovtcharova, J.: Towards higher configuration management maturity. IFIP Adv. Inf. Commun. Technol. 409, 385–395 (2013).

S55. Paulk, M., Curtis, W., Chrissis, M.B., Weber, C.: Capability Maturity Model for Software (Version 1.1). Software Engineering Institute, Carnegie Mellon University, Pittsburgh, PA (1993).

S56. CMMI Product Team: CMMI for Acquisition, Version 1.3. Software Engineering Institute, Carnegie Mellon University, Pittsburgh, PA (2010).

S57. CMMI Product Team: CMMI for Development, Version 1.3. Software Engineering Institute, Carnegie Mellon University, Pittsburgh, PA (2010).

S58. CMMI Product Team: CMMI for Services, Version 1.3. Software Engineering Institute, Carnegie Mellon University, Pittsburgh, PA (2010).

S59. Curtis, W., Hefley, W., Miller, S.: People Capability Maturity Model (P-CMM), Version 2.0. Software Engineering Institute, Carnegie Mellon University, Pittsburgh, PA (2009).

S60. The SGMM Team: Smart Grid Maturity Model, Version 1.2: Model Definition. Software Engineering Institute, Carnegie Mellon University, Pittsburgh, PA (2011).

S61. Duarte, A., da Silva, M.M.: Cloud Maturity Model. 2013 Ieee Sixth Int. Conf. Cloud Comput. Cloud 2013. 606–613 (2013).

S62. Introna, V., Cesarotti, V., Benedetti, M., Biagiotti, S., Rotunno, R.: Energy Management Maturity Model: an organizational tool to foster the continuous reduction of energy consumption in companies. J. Clean. Prod. 83, 108–117 (2014).

S63. Minonne, C., Turner, G.: Evaluating Knowledge Management Performance. Proc. 6th Int. Conf. Intellect. Cap. Knowl. Manag. Organ. Learn. 201–210 (2009).

S64. Visconti, M., Villarroel, R.: Managing the improvement of SCM process. In: Oivo, M. and KomiSirvio, S. (eds.) Product Focused Software Process Improvement, Proceedings. pp. 35–48 (2002).

S65. Rios, E., Bozheva, T., Bediaga, A., Guilloreau, N.: MDD maturity model: A roadmap for introducing model-driven development. In: Rensink, A. and Warmer, J. (eds.) Model Driven Architecture - Foundations and Applications, Proceedings. pp. 78–89 (2006).

S66. Salleh, H., Alshawi, M., Sabli, N.A.M., Zolkafli, U.K., Judi, S.S.: Measuring readiness for successful information technology/information system (IT/IS) project implementation: A conceptual model. Afr. J. Bus. Manag. 5, 9770–9778 (2011).

S67. Chong, L.M., Jing, Z.: Research on Agricultural and Sideline Products Logistics Capability Maturity Model. Proc. 2014 Int. Conf. Mechatron. Control Electron. Eng. 113, 727–731 (2014).

S68. Barclay, C.: Sustainable security advantage in a changing environment: The Cybersecurity Capability Maturity Model (CM2). Proc. 2014 Itu Kaleidosc. Acad. Conf. Living Converg. World Impossible Stand. (2014).

S69. Cusick, K., Minnich, I.: 6.2.3 Industrial Collaboration Systems Engineering Capability Maturity Model Description and Overview of Hughes Pilot Appraisal. INCOSE Int. Symp. 5, 634–641 (1995).

S70. McIntyre, M.H.: The Integrated Product Development Capability Maturity Model (CMM) SM. INCOSE Int. Symp. 6, 574–576 (1996).

Process mining applications in software engineering

Brian Keith[1] and Vianca Vega[1]

[1] Departamento de Ingeniería de Sistemas y Computación
Universidad Católica del Norte, Chile
{brian.keith, vvega}@ucn.cl

Abstract. Process mining is a field that uses elements from data mining and business process modeling to do tasks such as process discovery, conformance checking, and process improvement. This paper presents a study about the application of process mining techniques in the software development process. It shows a series of case studies that illustrate possible applications in the process and the product. Also, the main current challenges in applying process mining in software engineering are described. The objective of this paper is to show the importance and practical usefulness of applying process mining approaches in software engineering. The main result of this study is the fact that using process mining facilitates software process evaluation and auditing. The development of a methodology for applying process mining in software engineering is proposed as future work, considering the main challenges described previously.

Keywords: process mining, software development, process assessment, process improvement.

1 Introduction

Process mining is a discipline that lies between machine learning and data mining, and, process modelling and analysis. Its main objective is to discover processes, do conformance checking, and process improvement. Process mining seeks automatizing these three tasks by applying data mining techniques specially designed for dealing with process data [1-2].

Analyzing a process automatically and determining where deviations are produced may be used for taking corrective actions on how processes are done. This corresponds to the process improvement phase, where corresponding improvements are proposed, as a function of the information obtained [1].

Cook & Wolf in 1998 [3] intended to build software process models from data obtained from event records. Since then, process mining has greatly evolved and has become a field in itself, currently having applications beyond software engineering. Several studies applied in software engineering have been conducted, some of which are shown in the references [4-16].

Process mining applied in software engineering aims to solve monitoring and control problems and improving development processes, one of the critical problems in software development [17]. In general, an improvement in the quality of the

© Springer International Publishing AG 2017
J. Mejia et al. (eds.), *Trends and Applications in Software Engineering*, Advances in Intelligent Systems and Computing 537, DOI 10.1007/978-3-319-48523-2_5

development process is expected to result in a better product quality. This reinforces the usefulness of applying process mining techniques in this field since it would eventually allow developing higher quality software.

Another relevant aspect to consider is that software processes are not usually explicitly modelled. Manuals supporting development work contain guides and abstract procedures. Therefore, there is a big gap between the "actual" process and the "official" process [17-18]. This gap may be measured with process mining techniques, allowing correction and improving the process if necessary.

According to the above and under the perspective that software processes may be considered as software in themselves [19], the application of process mining techniques could be seen as equivalent to the application of automatized testing methods in software development.

Based on the review of process mining studies in software engineering, two main application dimensions are observed: in the process (process evaluation and improvement [3, 4, 6, 9, 11, 12, 13, 14, 16] and process support [5, 7, 15]) and in the product [8, 10].

The purpose of this study is to examine process mining applications in software engineering. Some studies review and analyze the use of data mining techniques in software engineering as a whole (e.g., the systematic review in [20]); however, none of these literature reviews focuses on the application of process mining techniques in existing software processes.

This paper is organized as follows: first, the basic concepts of process mining are explained. Then, a review of the main process mining applications in software engineering is presented. Next, some specific applications are described in detail as case studies, analyzing their main results. Afterwards, some challenges process mining faces in this field are explained. Finally, the main results are summarized and the conclusions obtained in this study are shown.

2   Process Mining and Its Application in Software Engineering

The basic task of process mining is process discovery. This task begins with a dataset consisting of a record of all events occurring during the process development. With a sufficient amount of these data, it is possible to infer the process model, with adjustable levels of detail [1].

Process discovery automatization enables organizations to eliminate the problem of process modelling conducted by experts because, in many cases, assessments are influenced by the modeler's own bias [26]. In addition, existing algorithms allow obtaining greater levels of detail or more abstract models omitting certain cases, according to the desire of the final user of the model [1]. Certainly, the quality of the resulting process models is a function of the quality and quantity of available data [22]. Given the nature of available data, the disadvantage is that the algorithm only includes examples that have occurred, but it does not have examples of processes that cannot occur. The lack of negative examples is one of the weaknesses of these approaches [1].

Conformance checking begins with a defined formal process model and an instance of the actual process, and generates an analysis of the actual process deviation as compared to the ideal process [23]. This is very useful for auditing and has been successfully applied previously in different organizations [24]. Conformance checking allows determining how faithfully the process model is followed with respect to its formal definition. There are several methods to do this task. All of them allow defining the level of deviation of an actual process with respect to the formal process and can determine in what step of the process deviation occurs. This is applied in process control and auditing of organizations [1, 24].

Finally, process improvement is based on the results obtained from the previous tasks. It begins with the actual process model and a performance indicator, generating a new process model which should improve results as compared to the original model [1]. A simple example is the elimination of a bottleneck detected via process discovery. It is possible to use simulation techniques to analyze possible improvements proposed and compare them according to a well-defined metric [25].

Process mining has been successfully used in various application domains [27-31]. There is an IEEE process mining community including more than 60 organizations [4]. However, despite process mining success, a limitation is that techniques are seldom used in operational environments. It has been shown that it is possible to use process mining for supporting operational decisions [31].

The big amount of data generated by the software development process allows applying data analysis techniques and using its results to guide process optimization. There are several examples of data mining techniques for software process analysis and improvement [20].

In order to apply process mining techniques, data available must fit the metamodel described in Fig. 1. Each process consists of activities and is associated with a series of process instances. Each process instance consists of one or more events. These events must indicate the activity they correspond to and have a description, a time to be executed, and a person in charge. It is possible to enrich the event record with more data, as available [16].

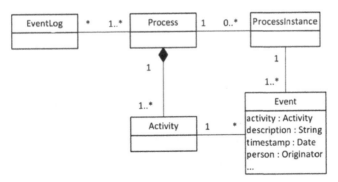

Fig. 1. Process mining metamodel [16]

Models may be worked out through different formal notations, from lower level models corresponding to transition diagrams to more advanced notations such as Petri networks, BPMN or UML [32].

Table 1 shows the dimensions of possible applications of process mining applications in software engineering.

Table 1. Process mining applications in software engineering

Dimension	Application	Example	References
Process	Process evaluation and improvement	Finding discrepancies between the formal and actual process.	[3], [4], [6], [9], [11], [12], [13], [14] & [16]
	Process support	Process discovery as a complement to requirement elicitation.	[5], [7] & [15]
Product	In the product	Analyzing users' behavior.	[8] & [10]

Fig. 2 illustrates some of the possible process mining applications in the context of software development. The life cycle presented is generic and follows the phases of the Rational Unified Process (RUP) [33].

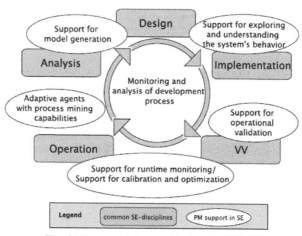

Fig. 2. Process mining applications in RUP [5]

In software engineering, there are several sources from which data could be obtained (e.g., version control repositories, bug tracking systems, and e-mail). Owing to the many available sources, a powerful step to prepare data is required before applying data mining procedures [16].

The broad applicability of process mining in software engineering is due to its generic techniques. These can be applied in any type of process, if correct data are available. Process mining can also be used for mining the use of software products developed. A case study analyzes two industrial applications, particularly a reservations system and a travel portal [8].

There are methodological proposals to apply process mining in software engineering, along with frameworks particularly designed to facilitate these tasks [9, 16, 35, 36]. Nevertheless, the development of a standard methodology is still an open problem in this area. This is due to the fact that software process intelligence is a very recent field still evolving [36].

3   Case Studies

This section shows the results of some studies in the literature. These studies were selected on the basis of their illustrative value. The case studies have been classified according to the process involved. Only the main results are shown here, in-depth detail of the implementations and modeling can be found in the referenced works.

Configuration management process. SCM data can be used to build explicit process models. This corresponds to an example of process discovery [9]. Using the audit data generated by the configuration management system, which correspond to process instances, the model can be derived from them using process mining algorithms. This model may be verified by a process engineer, who must later decide what changes must be made to optimize and manage it better [9].

By complementing process data with the person in charge of its activity, it is possible to build a model of the social network of the process and also analyze individuals who do each activity [1]. The application of these techniques reveals that, in some cases, people in charge of code development are the same individuals who are later in charge of its testing. This is certainly not desirable because, in ideal situations, they should be different people [9].

Testing process. It is possible to apply process mining techniques to obtain models from testing scripts used during the testing phase of software development. This allows generating new testing cases and facilitates understanding the different testing cases used [7]. Another study proposes the use of process mining techniques to analyze software systems online under real circumstances. The objective is to learn as much as possible from information obtained from these data in execution time and then use this knowledge to diagnose problems and recommend actions [10].

The first study [7] is an example of a process mining application as a supporting tool in the process because the purpose is to support the testing phase, but not to analyze the process. The second study [10] is an example of process support and also an application in the product because process mining tools are directly applied in the software to support the testing phase.

Change control process. This study fully describes the application of process mining techniques in data generated by the change control process in charge of a Change Control Committee in the context of a software development project [4].

One of the main results obtained revealed that in most process instances (about 70%) the analytical phase was omitted and change requests were directly solved without a previous analysis. This indicates that the omission of the analytical task is a structural problem in this organization [4]. This type of discrepancy must be detected automatically by applying process mining because it provides relevant knowledge about how processes operate in the development processes of an organization.

CMMI conformity assessment. This case corresponds to the application of conformity review techniques to evaluate a process according to CMMI [12, 13]. Studies have been previously done to assess conformity as compared to other standards. For example, an Australian software development company was assessed by taking standard ISO/IEC 12207 as a reference in 2006 [38].

CMMI model describes the typical elements processes must include to be effective [37]. In general, CMMI assessment consists of two steps: collecting data and then interpreting them according to the CMMI model [39]. Process assessment can be qualitative or quantitative, and it may rely on expert judgment [40]. Process mining may help in the data collecting phase. The use of automated process mining tools can help create objective assessments by reducing human intervention [1, 13].

Another aspect to which process mining can contribute is CMMI assessment cost reduction. One of the limitations of small environments in the application of CMMI process assessment is its high cost. The same study proposes a tool based on process mining which, once configured, constantly assesses the conformity of development processes with CMMI. This tool, called Jidoka4CMMI, provides a framework that defines testing cases and verifies if a process fits CMMI recommendations [13].

Another study applied process mining techniques to evaluate the real process done by developers in an organization. Then, discrepancies found between the actual process and the model described were assessed. The results of this assessment became a plan to work and improve the process [14].

4 Challenges

Some of the general challenges faced by process mining in the field of software engineering are presented below [1, 2, 13, 17, 22, 40, 41, 44].

Process knowledge. In many companies, information systems are agnostic to business processes. Actions done by the systems are kept in some kind of record archive, but the systems do not identify the process or process instance they correspond to [1, 41]. The problem of applying process mining in records generated by process-agnostic systems is one of the main challenges faced by process mining when applying it in the entrepreneurial world. Some of the problems emerging during data acquisition can be solved by defining and using standards such as MXML [42] or OpenXES [43], both based on XML [41].

In the case of process-agnostic systems, their records may not have enough data to apply process mining algorithms. In some cases, the opposite may occur since the amount of details is excessive and the complexity of the resulting model makes it impractical. Every good model must keep a balance between precision and the representation of reality and computing simplicity [2, 41].

Data quality. Process mining strongly depends on the quality of data collected and stored. Big amounts of data must be frequently filtered due to incomplete process instances. Nevertheless, useful datasets can be obtained through careful data preparation, although they were not data originally stored to apply process mining. In spite of this, process mining efficiency and effectiveness can be substantially benefited from well-structured and well-defined datasets and also collection guides that allow obtaining high-quality datasets [17, 22].

Another problem in datasets is noise. This may sometimes arise due to unexpected processes (activities which are not supposed to occur or occurring in incorrect positions) [41]. The integrity and quality of the data is an important factor in the quality of process assessment results [40], this becomes even more important in conjunction with the natural necessities of process mining techniques.

Tool integration. All software process improvement methodologies share a common pre-requirement: the process and the product must be constantly measured. To do this, development process data must be continuously collected and analyzed. Nevertheless, manual data collection and analysis require an important additional effort [13].

A key requirement for software process follow-up is that they should not interfere with the process under assessment. This is seldom the case for traditional approaches because they require the active participation of all process actors so that they can report their activities. An alternative approach is automatic data collection in work environments (e.g., IDEs and design tools). The use of these integrated tools may facilitate data obtainment to apply process mining techniques without interfering with software developers' work [13].

Result evaluation. Another problem is the assessment of the resulting process because it is necessary to define a standard and rigorous procedure to assess the quality of the output generated by the process mining algorithm [1]. It is possible to measure the quality of the model through the discrepancy between the records of original events and the model obtained. Another problem is to choose an adequate model representation because, ultimately, the model will not be useful if it is not possible to understand and analyze it properly [41].

Another challenge for the implementation of process mining techniques is that algorithms must be properly parametrized. The search of optimum parametrization can be a complex task due to the number of available parameters [1, 41].

Usability. According to the Process Mining Manifesto [2], one of the challenges that must be faced to improve process mining usability is its integration with other methodologies and analytical techniques. For example, simulation tools may complement process analysis by assessing alternative implementations [44].

In general, a redesign project requires more than one instrument. Although process mining provides tools for diagnosing and analyzing processes, they must be complemented with other methodologies (e.g. simulation) or process improvement tools to allow understanding and planning the redesign process properly [1, 2, 44].

5 Conclusions

This study addresses the main aspects necessary to understand the importance of process mining and its application in the field of software engineering. In addition, the different challenges and problems faced by process mining are identified. On the other hand, the study shows that process mining may be applied in the development process and also in the product developed. Several applications illustrating these concepts are shown. Given the general character of process mining algorithms, it is possible to apply them in almost all entities following a sequence of logical steps if data are in a proper format.

Process mining is an area with a great potential of applications in the field of software engineering, particularly in the study of software development processes.

Since process mining is a relatively new field as compared with the disciplines forming it, its growth potential is quite high, both on a theoretical and practical basis.

The capacity to discover actual processes, assess its conformity with respect to the official model, and find improvement opportunities, make process mining cause a high impact on an organizational basis. Case studies show that it is possible to use process mining to detect process discrepancies, obtaining many data that can be used for planning eventual corrections and improvements.

The challenges that must be faced in the field of process mining are also described. One of the main challenges is the availability of suitable data provided by information systems aware of business processes. This idea may be generalized to continuously assess process performance. This may allow managers to do a better job since they can be warned about possible process deviations and problems appearing in the real-time process in due time.

Although there are methodologies to apply process mining techniques in general, the development of a standard methodology to implement process mining techniques in organizations is still an open problem. Particularly, the need of a methodology is considered to explain the development of an activity to incorporate process knowledge to information systems naturally agnostic to it. Once this is done, tasks would somehow follow the traditional data mining approach, for which several methodologies are available.

Finally, the importance of process mining and automatic approaches of process assessment are shown both on a general basis and in software development processes.

References

1. W. Van Der Aalst, "Process mining: discovery, conformance and enhancement of business processes". Springer Science & Business Media, 2011.
2. W. Van Der Aalst, A. Adriansyah, A. K. A. de Medeiros, F. Arcieri, T. Baier, T. Blickle, J. C. Bose, P. van den Brand, R. Brandtjen, J. Buijs et al., "Process mining manifesto," in Business process management workshops. Springer, 2011, pp. 169–194.
3. J. E. Cook and A. L. Wolf, "Discovering models of software processes from event-based data", ACM Transactions on Software Engineering and Methodology (TOSEM), vol. 7, no. 3, pp. 215–249, 1998.
4. J. Šamalíková, J. J. Trienekens, R. J. Kusters, and A. T. Weijters, "Discovering changes of the change control board process during a software development project using process mining," in Software Process Improvement. Springer, 2009, pp. 128–136.
5. L. Cabac and N. Denz, "Net components for the integration of process mining into agent-oriented software engineering," in Transactions on Petri Nets and Other Models of Concurrency I. Springer, 2008, pp. 86–103.
6. A. Colombo, E. Damiani, and G. Gianini, "Discovering the software process by means of stochastic workflow analysis," Journal of Systems Architecture, vol. 52, no. 11, pp. 684–692, 2006.
7. V. Shah, C. Khadke, and S. Rana, "Mining process models and architectural components from test cases," in Software Testing, Verification and Validation Workshops (ICSTW), 2015 IEEE Eighth International Conference on. IEEE, 2015, pp. 1–6.
8. V. A. Rubin, A. A. Mitsyuk, I. A. Lomazova, and W. M. van der Aalst, "Process mining can be applied to software too!" in Proceedings of the 8th ACM/IEEE International Symposium on Empirical Software Engineering and Measurement. ACM, 2014, p. 57.

9. V. Rubin, C. W. Günther, W. M. Van Der Aalst, E. Kindler, B. F. Van Dongen, and W. Schäfer, "Process mining framework for software processes," in Software Process Dynamics and Agility. Springer, 2007, pp. 169–181.

10. W. v. d. Aalst, "Big software on the run: in vivo software analytics based on process mining (keynote)," in Proceedings of the 2015 International Conference on Software and System Process. ACM, 2015, pp. 1–5.

11. R. Santos, T. C. Oliveira et al., "Mining software development process variations," in Proceedings of the 30th Annual ACM Symposium on Applied Computing. ACM, 2015, pp. 1657–1660.

12. J. Samalikova, R. J. Kusters, J. J. Trienekens, and A. Weijters, "Process mining support for capability maturity model integration-based software process assessment, in principle and in practice," Journal of Software: Evolution and Process, vol. 26, no. 7, pp. 714–728, 2014.

13. S. Astromskis, A. Janes, A. Sillitti, and G. Succi, "Continuous CMMI assessment using non-invasive measurement and process mining," International Journal of Software Engineering and Knowledge Engineering, vol. 24, no. 09, pp. 1255–1272, 2014.

14. M. Huo, H. Zhang, and R. Jeffery, "An exploratory study of process enactment as input to software process improvement," in Proceedings of the 2006 international Workshop on Software Quality. ACM, 2006, pp. 39–44.

15. W. M. Van der Aalst, "Business alignment: using process mining as a tool for delta analysis and conformance testing," Requirements Engineering, vol. 10, no. 3, pp. 198–211, 2005.

16. Poncin, Wouter; Serebrenik, Alexander; Van Den Brand, Mark. "Process mining software repositories," in Software Maintenance and Reengineering (CSMR), 2011 15th European Conference on. IEEE, 2011. pp. 5-14.

17. J. Samalikova, R. Kusters, J. Trienekens, T. Weijters, and P. Siemons, "Toward objective software process information: experiences from a case study," Software Quality Journal, vol. 19, no. 1, pp. 101–120, 2011.

18. W. S. Humphrey, "A discipline for software engineering". Addison-Wesley Longman Publishing Co., Inc., 1995.

19. L. Osterweil, "Software processes are software too," in Proceedings of the 9th international conference on Software Engineering. IEEE Computer Society Press, 1987, pp. 2–13.

20. M. Halkidi, D. Spinellis, G. Tsatsaronis, and M. Vazirgiannis, "Data mining in software engineering," Intelligent Data Analysis, vol. 15, no. 3, pp. 413–441, 2011.

21. W. M. van der Aalst, "Process mining in the large: A tutorial," inBusinessIntelligence. Springer, 2014, pp. 33–76.

22. R. Bose, R. S. Mans, and W. M. van der Aalst, "Wanna improve process mining results?" in Computational Intelligence and Data Mining (CIDM), 2013 IEEE Symposium on. IEEE, 2013, pp. 127–134.

23. W. Van der Aalst, A. Adriansyah, and B. van Dongen, "Replaying history on process models for conformance checking and performance analysis," Wiley Interdisciplinary Reviews: Data Mining and Knowledge Discovery, vol. 2, no. 2, pp. 182–192, 2012.

24. W. M. van Aalst, K. M. van Hee, J. M. van Werf, and M. Verdonk, "Auditing 2.0: using process mining to support tomorrow's auditor," Computer, vol. 43, no. 3, pp. 90–93, 2010.

25. W. Van Der Aalst, "Process mining: overview and opportunities," ACM Transactions on Management Information Systems (TMIS), vol. 3, no. 2, p. 7, 2012.

26. F. Gottschalk, W. M. van der Aalst, and M. H. Jansen-Vullers, "Mining reference process models and their configurations," in On the Move to Meaningful Internet Systems: OTM 2008 Workshops. Springer, 2008, pp. 263–272.

27. R. Mans, M. Schonenberg, M. Song, W. M. van der Aalst, and P. J. Bakker, "Application of process mining in healthcare–a case study in a dutch hospital," in Biomedical Engineering Systems and Technologies. Springer, 2008, pp. 425–438.

28. A. Mitsyuk, A. Kalenkova, S. Shershakov, and W. van der Aalst, "Using process mining for the analysis of an e-trade system: A case study,", p. 15, 2014
29. N. Trcka and M. Pechenizkiy, "From local patterns to global models: Towards domain driven educational process mining," in Intelligent Systems Design and Applications, 2009. ISDA'09. Ninth International Conference on. IEEE, 2009, pp. 1114–1119.
30. C. Zinski. "Internet Banking: The Trade Name Trend". Banking & Financial Services Policy. 2001.
31. W. M. van der Aalst, M. Pesic, and M. Song, "Beyond process mining: from the past to present and future", in Advanced Information Systems Engineering. Springer, 2010, pp. 38–52.
32. A. A. Kalenkova, W. M. van der Aalst, I. A. Lomazova, and V. A. Rubin, "Process mining using BPMN: relating event logs and process models," Software & Systems Modeling, pp. 1–30, 2015.
33. P. Kruchten, "The rational unified process: an introduction". Addison-Wesley Professional, 2004.
34. M. Jans, M. Alles, and M. Vasarhelyi, "The case for process mining in auditing: Sources of value added and areas of application," International Journal of Accounting Information Systems, vol. 14, no. 1, pp. 1–20, 2013.
35. J. Stolfa, S. Stolfa, M. A. Kosinar, and V. Snasel, "Introduction to integration of the process mining to the knowledge framework for software processes," in Proceedings of the Second International Afro-European Conference for Industrial Advancement AECIA 2015. Springer, 2016, pp. 21–31.
36. A. Sureka, A. Kumar, and S. Gupta, "Ahaan: Software process intelligence: Mining software process data for extracting actionable information," in Proceedings of the 8th India Software Engineering Conference. ACM, 2015, pp. 198–199.
37. H. Glazer, J. Dalton, D. Anderson, M. D. Konrad, and S. Shrum, "CMMI or agile: why not embrace both!" 2008.
38. I. ISO, "IEC 12207 systems and software engineering-software life cycle processes," International Organization for Standardization: Geneva, 2008.
39. S. U. Team, "Standard CMMI appraisal method for process improvement (SCAMPI) a, version 1.3: Method definition document," 2011.
40. Bourque, Pierre, et al. "Guide to the software engineering body of knowledge (40 (R)): Version 3.0". IEEE Computer Society Press, 2014, ch. 8, sec. 3.2, p. 154.
41. A. Burattin, "Obstacles to applying process mining in practice", in Process Mining Techniques in Business Environments. Springer, 2015, pp. 59–63.
42. W. M. van der Aalst, B. F. van Dongen, C. W. Günther, R. Mans, A. A. De Medeiros, A. Rozinat, V. Rubin, M. Song, H. Verbeek, and A. Weijters, "Prom 4.0: comprehensive support for real process analysis", in Petri Nets and Other Models of Concurrency–ICATPN 2007. Springer, 2007, pp. 484–494.
43. C. W. Gunther, "XES standard definition". www.xes-standard.org, 2009, Cited on, p. 72.
44. S. Aguirre, C. Parra, and J. Alvarado, "Combination of process mining and simulation techniques for business process redesign: a methodological approach", in Data-Driven Process Discovery and Analysis. Springer, 2012, pp. 24–43.

Agile Practices Adoption in CMMI Organizations: A Systematic Literature Review

Marco Palomino[1], Abraham Dávila[2], Karin Melendez[3]
Marcelo Pessoa[4]

[1] Escuela de Posgrado, Pontificia Universidad Católica del Perú, Lima, Perú
palomino.marco@pucp.edu.pe
[2] Departamento de Ingeniería, Pontificia Universidad Católica del Perú, Lima, Perú
abraham.davila@pucp.edu.pe
[3] Departamento de Ingeniería, Pontificia Universidad Católica del Perú, Lima, Perú
kmelendez@pucp.edu.pe
[4] Polytechnic School, University of Sao Paulo, Sao Paulo, Brazil
mpessoa@usp.br

Abstract. In the recent years, the adoption of agile frameworks and methodologies in Software Development Organizations (SDO) has grown up considerably. Unfortunately, there are scenarios where agile practices can't cover alone all the needs; for instance, software projects with a large level of required formal documentation or large, complex software projects; similarly, this kind of situations happen frequently in a context of CMMI organizations. The aim of this study is identify and analyze the most used agile practices that are used in combination with CMMI within SDO. To accomplish this, a systematic literature review has been performed according to relevant guidelines. This study has identified practices related Daily Meeting and Product Backlog management from Scrum framework as most common agile practices used in combination with CMMI. In addition, we could identify that there are specific benefits of implementing practices from both approaches.

Keywords: Agile Practice, Agile Software Development, CMMI

1 Introduction

The methodologies or process models that have being used on the Software Development Organizations (SDO) have evolved over time; as a result, in the last years these organizations have considered (with more interest) the adoption of agile practices in software development [1], [2]. Regarding [1], [3], [4], this agile approach promotes a "fast" way of software development with product's partial deliveries.

On the other side, CMMI (Capability Maturity Model Integration), which is a model that groups best practices in development and maintenance activities [5], [6]; is a process model that has been adopted by many SDO [7], [8]. According to [9], the practices and process adoption from a certain level is relatively a challenge in small companies, that's why the importance of identifying agile practices which in concordance with CMMI could help in software development improvement's process.

© Springer International Publishing AG 2017
J. Mejia et al. (eds.), *Trends and Applications in Software Engineering*, Advances
in Intelligent Systems and Computing 537, DOI 10.1007/978-3-319-48523-2_6

The aim of this study is to identify agile practices commonly used in contexts of organizations which have already adopted CMMI. In fact, identifying most used agile practices in these kinds of organizations will allow recognizing activities and processes that could get higher benefits when implementing agile and CMMI together. To accomplish the goal of this research, a Systematic Literature Review (SLR) [10] was performed in the relevant digital databases.

The remainder of this paper is structured as follows: section 2 shows the background and related work; section 3, the methodology of the SLR; section 4, the results; and finally, in section 5, it is presented the final discussion and future work.

2   Background and Related Work

There are several studies related with agile practices and CMMI and how these different approaches work together First of all, there are researches of how agile practices could contribute to get CMMI maturity levels [5], [11]. Second of all, there are case studies [9], [14] that describe the consequences of implementing agile practices within an organization with CMMI culture.

On the other hand, in the Silva study [12] the authors analyzed the combined use of agile and CMMI through a SLR. This previous research [12] only considered studies published up to 2011 and the research questions were mainly focused on benefits and limitations of implementing both approaches. We could identify differences between their and our research. The newest in our research are: (i) verify if team's size affects the combined use of agile and CMMI, (ii) analyze studies and researches published up to 2016, which extends the scope of the previous research (iii) analyze if any agile methodology can be used in contexts of CMMI organizations. Due to these differences mentioned, we consider that this work is needed because it will include new primary studies and also, it will consider all recent agile practices incorporated in CMMI contexts.

3   Systematic Literature Review

3.1 Systematic Review Fundamentals

The research method used is a SLR based on the guidelines and lessons learned proposed by Brereton [13] and Kitchenham [10]. As part of Plan Review, it was specified four research questions.

In order to frame the research question and define the search string, it was used the PICOC (Population, Intervention, Comparison, Outcome, Context) criteria applied to software engineering [2]. The Table 1 shows the main keywords used on PICOC criteria and Table 2 shows the different Search Strings elaborated regarding all Data Sources used.

Table 1. Principal Keywords used based on PICOC Criteria.

Population:	Organizations with CMMI and Agile Methodologies
Intervention:	Agile Practice, Activity
Comparison:	None
Outcome:	Analysis, Researches
Context:	Software Engineering

Table 2. Search Strings

Data Source	Search String
Scopus	("CMMI" or "Capability Maturity Model Integration") AND ABS("Agil*" or "Agile Method" or "Agil* Software" or "Light*" or "Scrum" or XP or "Extreme Programming" or "Scrumban" or "Kanban" or "software development") AND("Practic*" or "Activit*" or "Software Development Practice" or "System Development" or "Application") AND("Research*" or "Mappin*" or "Evaluation*" or "Experience*")
IEEE Xplore	((("Abstract":(agile OR "Agile Methodology" OR scrum OR xp OR light* OR kanban)) AND "Abstract":cmmi OR "capability maturity model") AND "Abstract":software)
Elsevier ScienceDirect	ALL((cmmi or maturity)) and TITLE-ABSTR-KEY(("Agil*" OR "scrum" OR "xp" OR "light" OR "kanban"))[Journals(Computer Science)]
ACM Digital Library	content.ftsec:(cmmi "capability maturity model integration") AND (agile "software development" "Agil* Method*" scrum xp "extreme programming" kanban "light*")

On the other hand, the automatic search of primary studies was complemented by a manual search in the main repositories and conferences related with Agile and CMMI.

3.2 RSL Protocol

A RSL protocol was defined and adjusted later to reduce the possibility of researcher bias. This protocol was structured by six steps that included a first studies selection regarding the execution of Search String in scientific databases plus the original results obtained from the manual search. Then, the articles were analyzed, considering the article Title and Abstract and then the article Introduction and Conclusion. At the end, the final articles were verified by peer review in order to evaluate their exclusion or inclusion in our research.

The exclusion and inclusion criteria considered were:

- Inclusion criteria: Academic articles with methodological basis (mainly experiment, case study, Systematic Reviews, Systematic Mappings). In addition, only articles from sources mentioned in the research were considered. Also, we saw convenient to consider articles in Spanish and English language due to in recent years the agile approach in software development is widely adopted in Latin American companies. Finally, only articles that shows the combined use of agile and CMMI approaches were considered. Even if the article mentions agile practices, we don't use it unless the adoption of those practices is performed within an organization implementing CMMI.

In order to include articles that add significant value to our research, we also considered the reference list from all primary studies.

- Exclusion criteria: Duplicated articles were excluded and the search scope was limited to the following publication types: Journals, Conferences, Magazines, Technical Reports and Books. In addition, we excluded the articles that only show the results of adopting agile practices without considering CMMI contexts.

3.3 Quality Assessment

Quality Assessment of this SLR followed 11 criteria defined by [14] based on [15]. The following are the criteria used in the Quality Assessment:

- Is this study based on research?
- Is there a clear statement of the aims of the research?
- Is there an adequate description of the context?
- Was the study design appropriate to address the aims of the research?
- Was the selection strategy appropriate to the aims of the research?
- Was there a control group for comparing treatments?
- Was the data collected in a way that addressed the research aims?
- Was the data analysis rigorous enough?
- Has the relationship between researcher and participants been considered as an adequate degree?
- Is there a clear statement of results?
- Is the study relevant for practice or research?

According to [15], these mentioned criteria include three important issues related to quality, which were considered in the Quality Assessment:

- Rigor: a complete and adequate approach was applied to key research methods in the study?
- Credibility: are the results in a meaningful and well-presented way?
- Relevance: how useful are the results to the software industry and the scientific community?

For the assessment, each one of the primary studies obtained after inclusion and exclusion criteria of the RSL protocol was analyzed using the 11 questions defined. The scale used in the assessment had two values ("yes" or "no"). When the answer was affirmative, the criteria had a value of "1"; otherwise, the value was "0". As a result, the minimum value could be "0" and "11" as maximum value. In https://drive.google.com/drive/folders/0B_K_llz5juqPa2ZaRXRDbWFEZ28, there are the results of Quality Assessment.

3.4 Data Extraction and Data Synthesis Strategies

The Petersen Guides [16] suggest the exploration of some papers sections in case the abstract is not well-structured or vague. For this study and with the aim of answer all of the research questions, all the primary studies selected after last step of the RSL protocol were fully read. Then, the primary studies were grouped in order to associate it in a high level. In order to conduct the analysis, a narrative synthesis was defined [17]; especially the "Grouping and Clustering".

3.5 Studies Selection

The initial results are displayed in Table 3. Then, the duplicated studies were excluded using the list of all 2,375 studies. After that, the titles were revised in order to exclude irrelevant studies. After this step, 299 studies were selected.

Table 3. Data Sources of the Systematic Review

Type	Name of Database	Initial Results	Search Date
Automatic Search	IEEE Xplore	736	January, 2016
	ACM Digital Library	236	January, 2016
	Science Direct	215	January, 2016
	Scopus	1,078	January, 2016
Manual Search	Agile Journal, Agile Conferences and SEI Digital Library	110	January, 2016

Then, the abstracts of all 299 potential studies were reviewed in order to exclude the studies that don't consider agile practices and CMMI approaches. After abstract's review, a total of 75 studies were defined. Finally, the introduction and conclusion of all 75 studies were analyzed in order to get all the studies that considers agile and CMMI approaches. At this moment, 47 studies were identified. Finally, the references of the 47 studies were analyzed in order to get additional studies. At the end, 5 more studies were added and a total of 52 studies were identified.

4 Results

In addition to Research Questions, it was also performed a review of the 52 studies selected in order to analyze the publication years, publication channels and research types of all studies identified at the end of selection process. In (https://drive.google.com/drive/folders/0B_K_llz5juqPa2ZaRXRDbWFEZ28), there are the results obtained from the review. The following section will display the answer of the four Research Questions defined.

4.1 RQ-1. Why are agile practices implemented in organizations with CMMI culture?

First of all, we want to define if both approaches are compatible or not. From the analysis of the 52 primary studies, we could identify that both agile and CMMI approaches are not opposed to each other. In fact, both cultures share similar criteria and practices. From the previous premise, we can affirm that there is a compatibility level between agile and CMMI approaches which is corroborated in the studies [D1], [D3], [D4], [D5], [D6], [D7], [D9], [D12], [D14], [D16], [D17], [D20], [D23], [D28], [D30], [D32], [D34], [D35], [D37], [D44], [D45], [D48], [D49] and [D52] where we could identify that practices from different cultures, such as agile or CMMI, can be complemented each other in order to improve the current processes. Additionally, it proposes that the compatibility level is defined by the organizations. In conclusion, we can affirm that both approaches can coexist but there are some inconveniences mentioned in [D3], [D7], [D24], [D25], [D45] and [D27] that should be considered:

- Keep the agile principles while the agile practices and processes are extended.
- Identify the organization needs for a successful implantation of both approaches.
- Keep the premise that both approaches are complemented each other and there are no practices substituted by others.

Additionally, there are scenarios where the use of agile practices in combination with CMMI is appropriate; for instance, in the studies [D13], [D27] and [D29] there are situations where is recommendable getting certain level of flexibility and agility, which is achieved with agile practices adoption. Finally, we can conclude from the analysis of all 52 primary studies that agile practices are implemented in CMMI contexts because that combination allows the organizations to:

- Reduce the waste time inside the team
- Reduce the delivery time of the products
- Increase the team productivity
- Improve the competitiveness of organizations and product's quality
- Include flexibility and agility in the processes of CMMI organizations
- Improve communication with stakeholders using agile practices

4.2   RQ-2. Could any agile practice be used in combination with CMMI?

From the analysis of 52 primary studies we could identify that agile methodologies and practices are characterized mainly because they obey entirely to the agile principles and as long as this principles are respected, as we can see in the studies [D2], [D22] and [D33], any kind of agile practice could be adapted to any context, even those where the organization culture is more traditional.

Regarding agile practices using in CMMI contexts, we identify that in the studies [D3], [D4], [D5], [D6], [D7], [D8], [D9], [D10], [D12], [D16], [D17], [D19], [D20], [D22], [D23], [D24], [D25], [D26], [D29], [D30], [D32], [D33], [D34], [D35], [D36], [D37], [D39], [D41], [D43], [D44], [D45], [D47], [D48], [D50], [D51] and [D52], there are some agile practices mentioned. In addition, we could found that in the majority of the studies, Scrum and XP are mentioned. In fact, regarding [D8], [D20], [D39], [D43] and [D52] studies, both methodologies can be complemented each other.

In addition, there are also scenarios where the use of agile practices are not recommended; for instance, those scenarios where the kind of project requires an in-depth documentation and those where it is necessary to record all changes periodically. This point was evident in [D17], [D28] and [D52], which using agile practices that demand approach to customers is only possible according to the facilities that customers can give.

Finally, from the analysis of 52 studies, we could find that the agile practices applicability is defined by the organization's needs, when the needs are defined correctly and agile principles are considered, there are no restrictions. In Table 4, there are the main agile practices which are mentioned in the primary studies.

4.3   RQ-3. Is there any influence from the team's size in the agile practices use with CMMI culture?

The agile approach is distinguished due to, among others, it proposes intensely the interaction of team members; in fact, the success of the applicability of these practices is based on trust and compromise reached within teams [D47]. This feature negatively influences in the correct functioning of agile practices in scenarios where teams are larger, since as the team grows, it requires more robust and stable channels to allow free flow of communication. That's why in [D17], [D36], [D29] and [D32] are mentioned, for example, that small teams are ideal for implementing agile practices.

Additionally, [D48] and [D44] studies mention an important factor besides the team size, this factor is the location of the team members, and it is recommended that teams that implement agile practices should be in the same location because it requires intense communication and interaction among members. In contrast, CMMI provides an organizational infrastructure that allows successful projects with distributed teams, these types of teams can negatively influence in the adoption of agile practices in CMMI contexts.

Table 4. Agile practices mentioned in primary studies

Agile practice	Primary Studies
Daily Meeting	[D1], [D4], [D5], [D7], [D8] [D33], [D34], [D52]
Burndown Charts	[D2], [D4], [D13], [D14], [D24] [D28], [D52]
Story Points	[D13], [D21], [D28], [D36]
Sprint Meetings	[D4], [D17], [D24], [D28] [D30] [D43] [D45], [D52]
Retrospectives	[D13], [D21], [D24], [D35] [D43] [D45]
Backlog Management	[D7], [D8], [D12], [D13], [D17] [D21], [D24], [D28], [D34] [D43], [D45]
Continue Integration	[D8], [D34], [D39], [D50]

4.4   RQ-4. Are there primary studies related with the combined use of agile practices and CMMI?

With this RQ we pretend to analyze the interest of the scientific community on the combined use of agile and CMMI practices. To answer this question the results obtained in the selection step were analyzed. As shown in Appendix A, they were 52 primary studies obtained after the selection process. This is a great sign that there is a widespread interest from industry, since it is a large number of studies to analyze in a SRL. Additionally, if we refer to the BQ-1, BQ-2 and BQ-3, we can see that the interest is not only recent; in fact, there is a constant interest over the last 10 years.

Moreover, the interest of the scientific community can also be seen in the numerous studies that refer to mappings between agile and CMMI practices in different maturity levels. In Table 5 there is a list of all studies that we could find mappings between CMMI and agile practices.

Table 5. Primary studies with agile and CMMI mappings

CMMI Maturity Level	Primary Studies
CMMI Level 2	[D4], [D6], [D9], [D27], [D28], [D29] [D32], [D41], [D47], [D50]
CMMI Level 3	[D8], [D14], [D27], [D29], [D34], [D39] [D47], [D50]
CMMI Level 4	[D21], [D29]
CMMI Level 5	[D21]

5   Conclusion and Future Work

It can be concluded that there is an interest from industry and scientific community regarding the integration of agile and CMMI approaches. Both were considered by the software industry as guidelines with opposite principles and, in some circumstances, incompatible; however, we have found in recent researches that both share the same goals and that may converge to contribute beneficially to the organizations.

This compatibility between agile and CMMI approaches can take the best of both cultures because it is recognized that agile guidelines provide flexibility that enables organizations to respond to the constant changes, particularly in the management of requirements; on the other hand, organizations with agile guidelines are benefit from CMMI because they incorporate good practices that add formality in organizational infrastructure. On the other hand, we could verify that there are studies that indicate it is possible to get certification in the first CMMI maturity levels through the use of agile practices. In addition, using various practices from different agile methodologies allows organizations to apply for even higher maturity levels of CMMI.

During the SLR, there were validations in the planning and the methodology used. These validations were performed by other members of the project but despite peer review and assurance of the methodological framework, we have considered situations that can influence on the results and conclusions obtained. The main identified threat was the selection bias, because research results are conditioned by the proper selection of primary studies. The omission of any of the studies that can contribute to research is one of the most important threats to take into consideration.

Finally, in this study, a SLR was conducted in order to analyze studies about combined agile and CMMI approaches. From the research of RSL, we could find that there could be a further work related to the empirical validation of what is stated in the analysis of 52 primary studies. In addition, there could be future work about the verification of advantages and disadvantages in the use of both agile and CMMI approaches; as well as, the review of more successful cases of organization that mixes both guidelines.

Acknowledgements. This work is framed within the ProCal-ProSer project funded by Innóvate Perú under contract 210-FINCYT-IA-2013 and partially supported by the Department of Engineering and the Grupo de Investigación y Desarrollo de Ingeniería de Software (GIDIS) from the Pontificia Universidad Católica del Perú.

References

1. Dahlem, P. Diebold y Marc, «Agile Practices in practice: a mapping study.,» de *18th International Conference on Evaluation and Assessment in Software Engineering*, New York, 2014.

2. T. Dingsøyr, «A decade of agile methodologies: Towards explaining agile software development.,» *Journal of Systems and Software,* vol. 85, nº 6, pp. 1213-1221, 2012.

3. A. Cockburn,*Agile Software Development,*Reading, Massachusetts:Addison–Wesley, 2002.

4. B. W. Boehm y R. Turner, *Balancing agility and discipline: a guide for the perplexed,* Addison-Wesley, 2003.

5. C. J. Salinas, M. J. Escalona y M. & Mejías, «A scrum-based approach to CMMI maturity level 2 in web development environments,» de *14th International Conference on Information Integration and Web-based Applications & Services*, 2012.

6. SEI, «CMMI for Development, Version 1.3, CMU/SEI-2010-TR-033,» SEI, 2010.

7. A. Marcal, B. de Freitas, F. Furtado Soares y A. Belchior, «Mapping CMMI Project Management Process Areas to SCRUM Practices,» *Software Engineering Workshop,* pp. 13-22, 6 March 2007.

8. K. Łukasiewicz y J. Miler, «Improving agility and discipline of software development with the Scrum and CMMI,» *Software, IET,* vol. 6, pp. 416-422, October 2012.

9. A. Omran, «AGILE CMMI from SMEs perspective,» de *Information and Communication Technologies: From Theory to Applications. ICTTA*, 2008.

10. B. Kitchenham y C. S., «Guidelines for Performing Systematic Literature Reviews in Software Engineering. Technical Report EBSE-2007-01,» 2007.

11. T. Kähkönen y P. & Abrahamsson, «Achieving CMMI level 2 with enhanced extreme programming approach,» de *Product Focused Software Process Improvement*, Springer Berlin Heidelberg, 2004, pp. 378-392.

12. F. S. Silva, F. S. F. Soares, A. L. Peres, I. M. d. Azevedo, A. P. L. Vasconcelos, F. K. Kamei y S. R. d. L. Meira, «Using CMMI together with agile software development: A systematic review,» *Information and Software Technology,* pp. 20-43, 2015.

13. P. Brereton, B. A. Kitchenham, D. Budgen, M. Turner y M. & Khalil, «Lessons from applying the systematic literature review process within the software engineering domain.,» *Journal of systems and software,* pp. 571-583, 2007.

14. T. Dybå y T. Dingsøyr, «Empirical studies of agile software development: A systematic review,» *Information and Software Technology,* pp. 833-859, 2008.

15. B. J. Shea, J. M. Grimshaw, G. A. Wells, M. Boers, N. Andersson y C. Hamel, «Development of AMSTAR: A measurement tool to assess the methodological quality of systematic reviews.,» *BMC Medical Research Methodology,* 2007.

16. K.Petersen, R.Feldt y S.Mujtaba,«Systematic mapping studies in software engineering,» *12th International Conference on Evaluation and Assessment in Software Engineering*, 2008.

17. J. Popay, H. Roberts, A. Sowden, M. Petticrew, L. Arai y M. Rodgers, «Guidance on the conduct of narrative synthesis in systematic reviews: A product from the ESRC Methods Programme,» Lancaster University, Lancaster, 2006.

Appendix A: Studies included in the review

[D1] Clark, Catherine. «Get to CMMI ML3 Using Agile Development Processes for Large Projects.» Agile Conference. 2011.

[D2] Abdel-Hamid, A. N., & Hamouda, A. E. D. «Lean CMMI: An Iterative and Incremental Approach to CMMI-Based Process Improvement.» Agile Conference (AGILE). IEEE, 2015. 65-70.

[D3] Weller, Ed. «"Agile and CMMI: Friend or Foe? A Lead Appraiser's View".» Agile Journal, 2013.

[D4] Potter, N., & Sakry, M. «Implementing SCRUM and CMMI together.» Agile Journal, 2009: 1-6.

[D5] McMahon, P. E «Taking an agile organization to higher CMMI maturity.» AgileJournal,2012:19-23.

[D6] Alegría, J. A. H., & Bastarrica, M. C. «Implementing CMMI using a combination of agile methods.» CLEI electronic Journal, 2006: 1-15.

[D7] Jakobsen, C. R., & Sutherland, J. «Scrum and CMMI going from good to great.» Agile Conference. IEEE, 2009. 333-337.

[D8] MaAppluytCMM de nivel 5.» Revistas del IEEE América Latina, 2005.

[D9] Paulk, M. C. «Extreme programming from a CMM perspective.» IEEE, 2001. 19-26.

66   M. Palomino et al.

[D10] Baker, S. W. «Formalizing agility, part 2: How an agile organization embraced the CMMI.» Agile Conference. IEEE, 2006. 8.

[D11] Kovacheva, T. «Optimizing Software Development Process.» EUROCON - International Conference on Computer as a Tool (EUROCON). IEEE, 2011. 1-2.

[D12] Sutherland, J., Jakobsen, C. R., & Johnson, K. «Scrum and CMMI Level 5: The Magic Potion for Code Warriors.» Agile Conference. IEEE, 2007. 272-278.

[D13] Aggarwal, S. K., Deep, V., & Singh, R. «Speculation of CMMI in Agile Methodology.» Advances in Computing, Communications and Informatics (ICACCI). IEEE, 2014. 226-230.

[D14] Anderson, D. J. «Stretching Agile to fit Cmmi level 3 the story of creating msf for Cmmi process improvement.» Agile Development Conference. IEEE, 2005.

[D15] Bos, E., & Vriens, C. «An agile CMM.» En Extreme Programming and Agile Methods - XP/Agile Universe 2004, 129-138. Springer Berlin Heidelberg, 2004.

[D16] Garzás, J., & Paulk, M. C. «A case study of software process improvement with CMMI-DEV and Scrum in Spanish companies.» Journal of Software: Evolution and Process, 2013: 1325-1333.

[D17] de Souza Carvalho, W.C. «A Comparative Analysis of the Agile and Traditional Software Development Processes Productivity.» Computer Science Society (SCCC). IEEE, 2011. 74-82.

[D18] Selleri Silva, F., Santana Furtado Soares, F. «A Reference Model for Agile Quality Assurance: Combining Agile Methodologies and Maturity Models.» Quality of Information and Communications Technology (QUATIC). IEEE, 2014. 139-144.

[D19] Omran, A.«AGILE CMMI from SMEs perspective.»3rd International Conference. ICTTA, 2008.1-8.

[D20] Torrecilla-Salinas, C. J., Sedeño, J., Escalona, M. J., & Mejías, M. «Agile, Web Engineering and Capability Maturity Model Integration: A systematic literature review.» Information and Software Technology, 2016: 92-107.

[D21] Cohan, S., y H. Glazer. «An Agile Development Team's Quest for CMMI® Maturity Level 5.» Agile Conference. IEEE, 2009. 201-206.

[D22] Santana Furtado Soares, F, y S. Romero de Lemos Meira. «An Agile Strategy for Implementing CMMI Project Management Practices in Software Organizations.» Information Systems and Technologies (CISTI). IEEE, 2015. 1-4.

[D23] Pikkarainen, M., & Mantyniemi, A. «An approach for using CMMI in agile software development assessments: experiences from three case studies.» SPICE. 2006.

[D24] Trujillo, M. M., Oktaba, H., Pino, F. J., & Orozco, M. J. «Applying Agile and Lean Practices in a Software Development Project into a CMMI Organization.» En Product-Focused Software Process Improvement, 17-29. Springer Berlin Heidelberg, 2011.

[D25] Miller, J. R., & Haddad, H. M. «Challenges Faced While Simultaneously Implementing CMMI and Scrum: A Case Study in the Tax Preparation Software Industry.» Information Technology: New Generations (ITNG). IEEE, 2012. 314-318.

[D26] Gandomani, T. J. «Compatibility of agile software development methods and CMMI.» Indian Journal of Science and Technology, 2013: 5089-5094.

[D27] Łukasiewicz, K., & Miler, J. «Improving agility and discipline of software development with the Scrum and CMMI.» Institution of Engineering and Technology, 2012: 416-422.

[D28] Diaz, J., Garbajosa, J., & Calvo-Manzano, J. A. «Mapping CMMI level 2 to scrum practices: An experience report.» En Software process improvement, 93-104. Springer Berlin Heidelberg, 2009.

[D29] Marcal, A. S. C., de Freitas, B. C. C., Furtado Soares, F. S., & Belchior, A. D. «Mapping CMMI project management process areas to SCRUM practices.» Software Engineering Workshop. IEEE, 2007. 13-22.

[D30] Jakobsen, C.R, y K.A. Johnson. «Mature Agile with a twist of CMMI.» Agile Conference. IEEE, 2008. 212-217.

[D31] López-Lira Hinojo, F. J. «Agile, CMMI®, RUP®, ISO/IEC 12207...: is there a method in this madness?» SIGSOFT Software Engineering Notes. 2014. 1-5.

[D32] Lina, Zhang, y Shao Dan. «Research on Combining Scrum with CMMI in Small and Medium Organizations.» Computer Science and Electronics Engineering (ICCSEE). IEEE, 2012. 554 – 557.

[D33] Morris, P. D. «The Perfect Process Storm: Integration of CMMI, Agile, and Lean Six Sigma.» Crosstalk, 2012: 39-45.

[D34] Bougroun, Z., Zeaaraoui, A., & Bouchentouf, T. «The projection of the specific practices of the third level of CMMI model in agile methods: Scrum, XP and Kanban.» Information Science and Technology (CIST). IEEE, 2014. 174-179.

[D35] Pikkarainen, M. «Towards a Better Understanding of CMMI and Agile Integration - Multiple Case Study of Four Companies.» En Product-Focused Software Process Improvement, 401-415. Springer Berlin Heidelberg, 2009.

[D36] El Deen Hamouda, A. «Using Agile Story Points as an Estimation Technique in CMMI Organizations.» Agile Conference. IEEE, 2014. 16-23.

[D37] Kähkönen, T., & Abrahamsson, P. «Achieving CMMI level 2 with enhanced extreme programming approach.» En Product Focused Software Process Improvement, 378-392. Springer Berlin Heidelberg, 2004.

[D38] Gazzan, M., Shaikh, A. «Towards bridging the gap between CMMI and agile development methodologies.» CAINE. ISCA, 2014. 299-304.

[D39] Torrecilla-Salinas, C. J., Sedeño, J., Escalona, M. J., & Mejías, M. «Mapping Agile Practices to CMMI-DEV Level 3 in Web Development Environments.» International Conference on Information Systems Development (ISD). 2014.

[D40] Tuan, N. N., & Thang, H. Q. «Combining maturity with agility: lessons learnt from a case study.» Fourth Symposium on Information and Communication Technology. ACM, 2013. 267-274.

[D41] Marcal, A. S. C., de Freitas, B. C. C., Soares, F. S. F., Furtado, M. E. S., Maciel, T. M., & Belchior, A. D. «Blending Scrum practices and CMMI project management process areas.» Innovations in Systems and Software Engineering , 2008: 17-29.

[D42] Leithiser, R., & Hamilton, D. «Agile versus CMMI-process template selection and integration with microsoft team foundation server.» 46th Annual Southeast Regional Conference on XX. ACM, 2008. 186-191.

[D43] Salinas, C. J. Torrecilla, M. J. Escalona, y M. Mejías. «A scrum-based approach to CMMI maturity level 2 in web development environments.» Proceedings of the 14th International Conference on Information Integration and Web-based Applications & Services (IIWAS). New York: ACM, 2012. 282-285.

[D44] Konrad, Mike, y Shane McGraw. «CMMI & Agile.» SEI Webinar. SEI Digital Library, 2008.

[D45] Boehm, B., Turner's, R., & Network, P. I. «Love and Marriage: CMMI and Agile Need Each Other.» SEI Digital Library, 2010.

[D46] Irrazabal, E., Vásquez, F., Díaz, R., & Garzás, J. «Applying ISO/IEC 12207:2008 with SCRUM and Agile Methods Software Process Improvement and Capability Determination, 169-180. Springer Berlin Heidelberg, 2011.

[D47] Turgeon., J. «SCRUMP (Scrum + RUP) and CMMI:The Story of a Harmonious Process and Product Deployment.» 2011.

[D48] Glazer, H., Dalton, J., Anderson, D., Konrad, M. D., & Shrum, S. CMMI or agile: why not embrace both! Technical Report, SEI, 2008.

[D49] Turner, R.,& Jain, A.«Agile Meets CMMI:Culture Clash or Common Cause?»—XP/Agile Universe 2002, 153-165.Springer Berlin Heidelberg, 2002.

[D50] Santana, C., Gusmão, C., Soares, L., Pinheiro, C., Maciel, T., Vasconcelos, A., & Rouiller, A. «Agile Software Development and CMMI: What We Do Not Know about Dancing with Elephants.» En Agile Processes in Software Engineering and Extreme Programming, 124-129. Springer Berlin Heidelberg, 2009.

[D51] Vriens, C. «Certifying for CMM Level 2 and ISO9001 with XP@ Scrum.» Agile Development Conference. IEEE, 2003. 120-124.

[D52] Mahnic, V., & Zabkar, N. «Introducing CMMI Measurement and Analysis Practices into Scrum-based Software Development Process.» International Journal of Mathematics and Computers In Simulation, 2007: 65-72.

Environmental Sustainability in Software Process Improvement: a Systematic Mapping Study

Gabriel Alberto García-Mireles

Departamento de Matemáticas, Universidad de Sonora
Blvrd. Encinas y Rosales s/n Col. Centro
83000 Hermosillo, Sonora, México
mireles@mat.uson.mx

Abstract. Sustainability is a main concern in our current society. One of the aspects that play an important role in supporting sustainable development is Information Technology (IT). Both software behavior and the way it is developed impact the amount of energy consumption. Thus, this paper aims to present the most recent approaches to address sustainability from a software process improvement perspective. A systematic mapping study was conducted to identify the latest efforts made in the IT field to improve sustainability. As a result, seven primary papers with initial ideas about how sustainability can be integrated into software processes were found. The lack of both proposals and empirical data suggests that further research on the topic is needed.

Keywords: Environmental sustainability, green software, software process improvement.

1 Introduction

Sustainability is a research area of recent interest among software engineering researchers [1] since energy consumption is one of the main concerns for IT. As hardware increases its power, the influence of software behavior on energy consumption grows too [2]. Usage scenarios can determine the energy consumption of servers and software applications. While energy consumption for the former can increase up to 40%, it can be of 20% for the latter [2]. Indeed, companies, governments and society in general have to deal with environmental issues and adopting environmental sound practices [3].

Environmental issues can also be treated as the green dimension of sustainable development [4]. Green software is an application that produces as little waste as possible during its development and operation [5]. Green software definition proposals address the topics of energy consumption and waste reduction during software product life cycle. A more formal definition of sustainable software is as follows: "software which direct and indirect impacts on economy, society, human being, and environment that result from development, deployment and usage of the software are minimal and/or which has a positive effect on sustainable development" [6].

In software engineering, the research on sustainable software is in its initial stage [7]. Naumann et al. [8] pointed out that it is important to investigate the meaning of

© Springer International Publishing AG 2017
J. Mejia et al. (eds.), *Trends and Applications in Software Engineering*, Advances in Intelligent Systems and Computing 537, DOI 10.1007/978-3-319-48523-2_7

sustainable software and sustainable software engineering. Practitioners, on the other hand, need practices, methods and tools to define and develop green software [9]. Sustainability is a systemic aspect that has multiple dimensions and requires actions in multiple levels [10].

In this systematic mapping study (SMS), the goal is to understand the extent to which sustainability is addressed in the software process literature in order to identify the main approaches on sustainable software developing research by either using an improved software process or improving the software process for enhancing product sustainability. In addition, the SMS can identify research gaps that can be addressed in future research projects.

The paper is structured as follows. Section 2 describes the relevant literature reviews about sustainability and software process. Section 3 depicts an overview of the main aspects considered during the planning stage of the SMS. Section 4 describes the main outcomes of the SMS while Section 5 presents the discussion of these results. Finally, conclusions and suggestions for future research on the field are addressed in Section 6.

2 Related Work

Two literature reviews on software engineering and sustainability were found. In the first one, conducted in 2012 [11], the authors reported that the first work on sustainability in software engineering was published in 2006 and that by 2012 only few papers had addressed the subject. As a conclusion, they stated that there was little methodological guidance to support sustainability.

In a second review, published in 2014, Penzendstadler et al. [1] found that software engineering process, software design and software quality were the main areas in Software Engineering that had addressed sustainability. They concluded that although sustainability had received widespread attention, little evidence about implementing sustainability proposals in industrial settings had been reported [1].

Other works were reviewed to understand how software process improvement is addressed in the literature and to identify relevant search terms. Terms similar to those by Garcia-Mireles et al. [12], Unterkalmsteiner et al. [13], and Pino et al. [14] were searched and analyzed.

3 Planning the Systematic Mapping Study

The main goal of this systematic mapping study is to identity the extent to which software process improvement field has addressed environmental sustainability. The general research question is:

- What approaches have been used to address environmental sustainability concerns in the software process improvement field?

Since environmental sustainability research area is a recent one, the main purpose of the research is to understand the main ideas explored in papers to improve software process taking into account sustainability goals. These ideas can be expressed as proposals or they may be empirically validated. The main research question is decomposed into specific questions to classify relevant literature as it is showed in Table 1.

Table 1. Specific research questions

Number	Question	Rationale
RQ1	What are the publication trends in regards to sustainability enhancement based on software process improvement?	To understand the extent to which the research topic is investigated
RQ2	What research approaches are reported in primary papers?	To uncover the opportunities to conduct empirical studies and contribute with new proposals
RQ3	What software processes have been improved?	To identify improved software processes
RQ4	What are the main features of identified approaches to improve sustainability in software processes?	To identify specific practices, methods, processes or frameworks, used to enhance sustainability

Selection criteria are as follows:
1. The article must be written in English and must be on databases until July 20, 2016.
2. The article must address the topic of environmental sustainability in the context of software process.
3. The article can include empirical data as well as conceptual models.
4. A full version of the article must be available.

Exclusion criteria are as follows:
1. The article addresses sustainability of software organization in an industrial context or sustainability of improvement approaches implemented in organizations, but they do not consider the environmental sustainability dimension.
2. The article presents a literature review about sustainability and software process. Literature reviews on sustainability and software engineering were previously searched.
3. The article is a forward, editorial, extended abstract, poster or thesis.

Table 2. Search string

Key terms	Synonyms
Software process improvement	"software process" OR "software process improvement" OR spi OR "process improvement" OR cmm OR cmmi OR 15504 OR spice OR 12207 OR 9001
Environmental sustainability	sustainab* OR green OR "energy efficient" OR ecolog*

The search string was derived from the literature reviews presented in the Related Work section. The Table 2 depicts the terms used in the searching process. These terms were used to carry out an automatic searching process on ACM Digital Library, IEEEXplore, Web of Science and Scopus databases.

4 Results

The search string was executed in four databases (Table 3) commonly used in Software Engineering research [15, 16]. Since each database has different interfaces to input search terms, the search string (see Table 2) was set to run in databases. A total of 414 records were retrieved until July 20th, 2016 (see Table 3). After reading the titles and/or abstracts of the papers, 25 candidates matched the selection criteria and their full versions were searched on databases. After reading them, seven of the works were classified as primary papers, like shown on Table 3.

Table 3. Selection of primary papers

Database	Retrieved	Candidates	Primary
ACM	12	1	0
IEEE	99	6	2
Web of Science	82	4	2
Scopus	221	14	3
Total	*414*	*25*	*7*

4.1 RQ1. Publication Trends

Publication trends show that since 2011 there is evidence that researchers have addressed the sustainability of software products taking into account the process view. Particularly, the article published in 2011 focused on the main aspects of the software life cycle stages that have potential impact on the environmental sustainability [17]. Others researchers have addressed software process models, such as ISO/IEC 12207 [18] to define sustainable software processes [19] or to define a sustainable software life cycle [20].

As Table 4 shows, research on sustainability in software process is in its initial stage. At least, one paper has been published per year from 2011 to 2015. The papers are published in conferences, journals and book chapters (see Fig. 1). As regards to specific software process conferences, a couple of papers were published on EuroSPI [19] and ICSSP [23]. Two more conference papers were published in measurement-oriented conferences. Journals such as Sustainable Computing: Informatics and Systems and International Journal of Software Engineering and Its Applications have published research about sustainability on software process. In summary, the topic is in its initial research stage and traditional conferences on software process have addressed this topic. Likewise, there is a specialized journal about sustainability that has also published works on software process.

Table 4. List of primary papers

Paper ID	Year	Reference	Paper goal
Lami2014	2014	[21]	Evaluates the software process sustainability
Lami2012	2012	[22]	Addresses the need of setting sustainability goals and measuring them
Lami2012S	2012	[19]	Provides a core set of processes to introduce environmental sustainability into organizations
Naumann2011	2011	[17]	Presents the GREENSOFT Model
Lami2013	2013	[23]	Provides a method to deriving sustainability metrics
Mahmoud2013	2013	[20]	Builds a two-level-green-software model that covers the sustainable life cycle of a software product and the software tools promoting green software
Naumann2015	2015	[8]	Sustainable software process models are examined

Source of primary papers

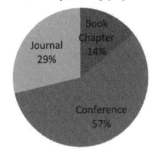

Fig. 1. Source-based classification of primary papers

4.2 RQ2. Research Type

Wieringa's [24] categorization was used to classify primary papers. After reading them, no evidence of empirical data that supports the proposals of the primary papers was found. Actually, the main ideas presented in them are about creating new models of sustainable software life cycles and new processes for supporting sustainability while software organizations develop software. Such ideas correspond to philosophical papers in Wieringa's [24] since this type of research paper describes a new way of looking at things or provides new conceptual frameworks.

4.3 RQ3. Software Processes Addressed

Three main approaches to address sustainability in software process improvement were identified. A general view discusses software life cycle stages and the way they can impact environmental sustainability [17, 20]. A second approach addresses processes oriented towards sustainability enhancement derived from models such as ISO/IEC 12207 [19]. The third approach focuses on improving specific processes that can be mapped to processes described into ISO/IEC 12207 [20].

Fig. 2 presents the main processes considered in primary papers. Lami et al. [21, 23] present proposals to measure sustainability, which is seen like the capability attribute of software process. They also address sustainability as a new quality characteristic of ISO/IEC 25010 [25] to develop a process for enhancing software product sustainability. Naumann et al. [17] discuss how sustainability can be taken into account when an organization needs to acquire a new software. On the other hand, Mahmoud et al. [20] discuss how a sustainable software life cycle can include specific activities in requirement process, design process and testing process. However, the description of these processes [20] do not address the requirements of title (process name), purpose and outcomes suggested to document software processes.

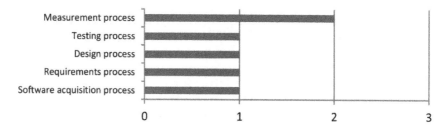

Software processes addressed in primary papers

Fig. 2. Processes from ISO/IEC 12207 addressed in primary papers

4.4 RQ4. Proposals Main Features

Three main approaches to address software processes are identified in primary papers: developing sustainability software life cycle model, creating a sustainability process reference model and developing specific processes focused on enhancing either software product sustainability or software process sustainability.

Developing sustainability software life cycle models relies on the work of Naumann et al [17], who presented the GreenSoft Model (Green and Sustainable Software Model). The model has two main goals: developing sustainable software and reducing both energy and resource consumption in IT. The model takes into account the three-order impacts of IT on the environment as well as the three main dimensions: environmental, social and economic. The model is organized into four central parts: holistic life cycle model that considers product life cycle from its conception to its disposal, sustainability criteria and metrics, procedure models for different stakeholders and tool support to carry out activities. The model includes four general activities related to a continuous improvement cycle: sustainability reviews, process assessment, sustainability journal and sustainability retrospective [17]. In Naumann et al. [8], on the other hand, authors provide a description to develop a sustainable agile software.

Based on GreenSoft model, Mahmoud and Ahmad [20] propose a software life cycle model that covers aspects of software related to green computing. The model has two levels that take into account sequential, iterative and agile practices to create a green and sustainable software process. The model includes both requirement and testing stages and describes guidelines to carry out activities in each software stage. The second level addressed how software can be used as a tool to promote green IT.

Creating a sustainability process reference model is a research strategy that develops new processes to address sustainability. Based on ISO/IEC 12207[18] and ISO/IEC 15504 [26], Lami et al. [19] define three processes: sustainability management process, sustainability engineering process and sustainability qualification process. These processes can support organizational sustainability goals. In addition, the paper provides a capability profile to evaluate the greenness of an organization.

The last research strategy identified was the development of enhanced process models to address environmental sustainability. This includes developing a measurement software process framework based on ISO/IEC 33000 series [27] in order to establish a sustainability model similar to capability models of software process [21]. In [21], authors define sustainable software process as a process that can meet established sustainability goals. In addition, the paper defines sustainability levels and process attributes for a sustainable process. Finally, the measurement process is adapted in order to address sustainability concerns [21].

5 Discussion

Sustainability research in the context of software process is in an initial stage. Despite studies on sustainability and software engineering inspired workshops since 2009 [10], software process has been a relevant issue since 2011 when GreenSoft model was published [17].

Software life cycle is a consistent topic within software process research [28]. Cugola et al. [29] noted that the concept of software life cycle is an initial solution to software development. The life cycle defines the standard life of a product and the development phase is decomposed into a predefined sequence of phases, as described by Naumann et al. [17]. Each phase is organized as a set of activities, as Mahmoud et al. [20] presented. However, further research is necessary to fully describe the software life cycle models considering appropriate software processes.

It was also found that although some specific processes are addressed in primary papers, the majority of them are not fully defined with the intention of being used as reference in a software organization. Therefore, a suitable approach to document these processes and the way sustainability is addressed is advised. In addition, guidelines to tailor the processes to specific organizational needs also have to be addressed.

As the results suggest, there is a lack of validation proposals in industrial settings. Sustainability is a concern that needs to be addressed in a systemic way and it imposes a different responsibility on each software project stakeholders [10], including software developers and project managers.

In summary, this SMS presents the trends on how sustainability is addressed within the software process improvement field. Even though the focus was on classifying research into a specific topic, solid procedures to get useful outcomes must be followed. Hence, guidelines for conducting systematic literature reviews were used [30]. However, bias in paper selection and extraction procedures are the main limitations of a systematic review [30, 31].

In order to mitigate the impact of selection bias, a protocol was built in the first stages of the research. In this paper, key search terms were identified and synonyms were included. All of the databases used were appropriate for conducting systematic reviews in software engineering [15] and only peer-reviewed articles, including conference proceedings, were selected [16]. However, papers that did not fulfilled this requirement were not included. Human error is another aspect that can impact paper selection. Thus, search and selection procedures were kept in a log to avoid potential issues. However, the paper selection process might be biased when there is only one person working on it [15].

Finally, to reduce bias due to extraction procedures, a template was built to extract verbatim data from each primary paper. Obtained data let classify papers and identify main approaches to deal with sustainability from a software process perspective.

6 Conclusions

In this systematic mapping study, seven primary papers addressing sustainability in a software process context were found. Since 2011, at least one paper has been published every year and research papers provide conceptual models for introducing sustainability during software development. Processes for measurement, acquisition, requirements, design and testing have been adapted to enhance sustainability. The primary papers reviewed considered both improving sustainability of a software product and improving sustainability of the software process.

Three main approaches have been recommended to deal with sustainability in the software process field: identifying the impact of each software life cycle stage on sustainable development, developing new processes to address sustainability as a product and process concern, and adapting current software process technology to include sustainability. Nevertheless, the lack of empirical data hinders these ideas to be applied in industrial settings.

This review provides a basis to study means to increase sustainability in software products by improving software process. As future work, it is possible to define detailed software processes for each software life cycle. In addition, guidelines to apply sustainable software processes on different settings are necessary. Furthermore, empirical studies on sustainability of current software processes in industrial settings need to be carried out and proposals to improve sustainability in software processes need to be validated.

References

1. Penzenstadler, B., Raturi, A., Richardson, D., Calero, C., Femmer, H., Franch, X.: Systematic mapping Study on Software Engineering for Sustainability (SE4S). In: ACM International Conference Proceeding Series, pp. 1-10. (2014)

2. Ardito, L., Procaccianti, G., Torchiano, M., Vetrò, A.: Understanding green software development: A conceptual framework. IT Professional. 17, 44-50 (2015)

3. Murugesan, S.: Harnessing Green IT: Principles and Practices. IT Professional. 10, 24-33 (2008)

4. Calero, C., Piattini, M.: Introduction to green in software engineering. In: Calero, C.Piattini, M., (eds.). Green in Software Engineering. pp. 3-27. (2015)

5. Erdélyi, K.: Special factors of development of green software supporting eco sustainability. In: SISY 2013 - IEEE 11th International Symposium on Intelligent Systems and Informatics, Proceedings, pp. 337-340. (2013)

6. Dick, M., Naumann, S.: Enhancing software engineering processes towards sustainable software product design. In: EnviroInfo, pp. 706-715. (2010)

7. Lago, P., Meyer, N., Morisio, M., Muller, H.A., Scanniello, G.: 2nd International workshop on green and sustainable software (GREENS 2013). In: Proceedings - International Conference on Software Engineering, pp. 1523-1524. (2013)

8. Naumann, S., Kern, E., Dick, M., Johann, T.: Sustainable software engineering: Process and quality models, life cycle, and social aspects. In: Hilty, L.M.Aesbischer, B., (eds.). ICT Innovations for Sustainability. 310, pp. 191-205. (2015)

9. Penzenstadler, B., Femmer, H.: A generic model for sustainability with process- and product-specific instances. In: GIBSE 2013 - Proceedings of the 2013 Workshop on Green in Software Engineering, Green by Software Engineering, pp. 3-7. (2013)

10. Becker, C., Chitchyan, R., Duboc, L., Easterbrook, S., Penzenstadler, B., Seyff, N., Venters, C.C.: Sustainability Design and Software: The Karlskrona Manifesto. In: Proceedings - International Conference on Software Engineering, pp. 467-476. (2015)

11. Penzenstadler, B., Bauer, V., Calero, C., Franch, X.: Sustainability in software engineering: A systematic literature review. In: IET Seminar Digest, pp. 32-41. (2012)

12. García-Mireles, G.A., Moraga, M.Á., García, F., Piattini, M.: Approaches to promote product quality within software process improvement initiatives: A mapping study. Journal of Systems and Software. 103, 150-166 (2015)

13. Unterkalmsteiner, M., Gorschek, T., Islam, A.K.M.M., Cheng, C.K., Permadi, R.B., Feldt, R.: Evaluation and Measurement of Software Process Improvement: A Systematic Literature Review. IEEE Transactions on Software Engineering. 38, 398-424 (2012)

14. Pino, F.J., García, F., Piattini, M.: Software process improvement in small and medium software enterprises: a systematic review. Software Quality Journal. 16, 237-261 (2008)

15. Kitchenham, B., Brereton, P.: A systematic review of systematic review process research in software engineering. Information and Software Technology. 55, 2049-2075 (2013)

16. Petersen, K., Vakkalanka, S., Kuzniarz, L.: Guidelines for conducting systematic mapping studies in software engineering: An update. Information and Software Technology. 64, 1-18 (2015)

17. Naumann, S., Dick, M., Kern, E., Johann, T.: The GREENSOFT Model: A reference model for green and sustainable software and its engineering. Sustainable Computing: Informatics and Systems. 1, 294-304 (2011)

18. Systems and software engineering -- Software life cycle processes - Redline. ISO/IEC 12207:2008(E) IEEE Std 12207-2008 - Redline1-195 (2008)

19. Lami, G., Fabbrini, F., Fusani, M.: Software sustainability from a process-centric perspective. In: Winkler, D., O'connor, R.V., Messnarz, R., (eds.). Systems, software and services process improvement. European Conference on Software Process Improvement. 301 CCIS, pp. 97-108. (2012)

20. Mahmoud, S.S., Ahmad, I.: A green model for sustainable software engineering. International Journal of Software Engineering and Its Applications. 7, 55-74 (2013)

21. Lami, G., Fabbrini, F., Buglione, L.: An ISO/IEC 33000-compliant measurement framework for software process sustainability assessment. In: Proceedings - 2014 Joint Conference of the International Workshop on Software Measurement, IWSM 2014 and the International Conference on Software Process and Product Measurement, Mensura 2014, pp. 50-59. (2014)
22. Lami, G., Buglione, L.: Measuring Software Sustainability from a Process-Centric Perspective. In: Software Measurement and the 2012 Seventh International Conference on Software Process and Product Measurement (IWSM-MENSURA), 2012 Joint Conference of the 22nd International Workshop on, pp. 53-59. IEEE, (2012)
23. Lami, G., Fabbrini, F., Fusani, M.: A methodology to derive sustainability indicators for software development projects. In: Procedings of the 2013 International Conference on Software and System Process, pp. 70-77. (2013)
24. Wieringa, R., Maiden, N., Mead, N., Rolland, C.: Requirements engineering paper classification and evaluation criteria: A proposal and a discussion. Requirements Engineering. 11, 102-107 (2006)
25. ISO/IEC-25010: Systems and software engineering - Software product Quality Requirements and Evaluation (SQuaRE) - Software product quality and system quality in use model. (2010)
26. ISO15504: ISO/IEC 15504 Information Technology -- process assessment -- part 3: guidance on performing an assessment. (2004)
27. ISO33001: ISO/IEC 33001:2015 Information technology -- process assessment -- concepts and terminology. (2015)
28. Bourque, P., Fairley, R.E., eds. Guide to the Software Engineering Body of Knowledge, Version 3.0. 2014, IEEE Computer Society.
29. Cugola, G., Ghezzi, C.: Software Processes: a Retrospective and a Path to the Future. Software Process: Improvement and Practice. 4, 101-123 (1998)
30. Kitchenham, B., Charters, S., Guidelines for performing systematic literature reviews in software engineering, in Technical report, Ver. 2.3 EBSE Technical Report. EBSE. (2007).
31. Dybå, T., Dingsøyr, T.: Empirical studies of agile software development: A systematic review. Information and Software Technology. 50, 833-859 (2008)

CAPE Role in Engineering Innovation: Part 1-The evolution

Luis Puigjaner

Universitat Politècnica de Catalunya Barcelona Tech,
Department of Chemical Engineering CEPIMA,
Diagonal 647, E-08028 Barcelona, Spain
luis.puigjaner@upc.edu

Abstract. The advancement of science in the past century gave rise to a number of revolutionary discoveries that deeply affected the way of life of our society. Here, is given a personal summary of the variety of applications evolving from a major discovery, the analytical engine, which instrumented a novel, revolutionary *software engineering*, enhancing the now so called Computer Aided Process Engineering in a variety of applications. The race among software-hardware has made possible fast pace in the *evolution/revolution* that affects all branches of science towards new discoveries with different impact and magnitude. The reader is guided on a tour through various milestones lived, whose main protagonist is an increasingly sophisticated software. Applications to a variety of *systems*, like telecommunications, biology, chemical engineering, mechanics, mining, etc. are revisited.

Keywords: Software Engineering, Computer aided process engineering, Process systems engineering.

1   Introduction

It has been fascinating to observe the advances made by science in the last century. We witnessed an incredible number of discoveries that changed our lifestyles, affected our society and had a profound effect on deeply held beliefs. Some of these discoveries were *revolutionary*. Television, atomic energy and satellite communications form part of a list of inventions too long to mention in full [1].

One significant discovery was the *difference engine*, which could be said to be the first computer. It was the computer pioneer Charles Babbage (1791-1871) who devised two classes of engines, the difference engine and the *analytical engine* [2]. This last one gave place to fully-fledged general-purpose computation leading to the now so called Computer Aided Process Engineering in a variety of applications. This *revolutionary* computing invention opened new doors and perspectives, mainly because it offered solutions to problems that had for a long time remained unsolved.

Let's go on to *describe some examples* that I have come across in my personal experience of CAPE's role in the evolution of engineering.

© Springer International Publishing AG 2017   79
J. Mejia et al. (eds.), *Trends and Applications in Software Engineering*, Advances in Intelligent Systems and Computing 537, DOI 10.1007/978-3-319-48523-2_8

2   Samples of the past

2.1   Modeling Telecommunication Systems

When I became part of NASA Aerospace Telecommunications Unit in 1965, the avalanche of great discoveries that surrounded us was fascinating. In my case, my quest was to find a way of conveying a huge number of signals with a minimum loss of power. Multiplexing systems (Fig. 1) already existed, but inter-channel crosstalk was a major problem in manned space flights. On these flights, trajectory and telemetry data, which provide both positional and environmental parameters about the spacecraft and the astronauts, must be processed immediately upon their arrival by a real-time operation system via a worldwide communications network. A new framework for a fast and reliable telecommunications system in real time was needed.

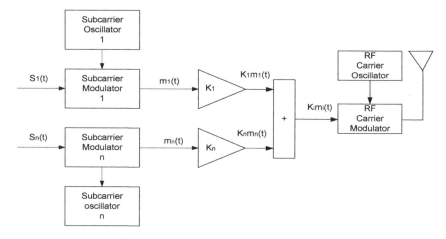

Fig. 1. Multiplexing System SSC-FM block diagram

The model proposed uses a novel analytical signal representation for the real signal. Its advantage is that the phase and envelope of the real signal can be fully described and represented in the upper half of the complex plane. Moreover, the phase and envelope are actually given in terms of the zeros of the analytical signal: the so-called "zero-locus" [3].

The phase or envelope can be manipulated to produce different signals that do not interfere with one another. Thus, a "common envelope set" will be obtained that contains signals that differ in phase but not in envelope or bandwidth. Computationally speaking, this means that one may easily track the real modulating signal through the zeroes of its analytical signal in the complex Z-plane.

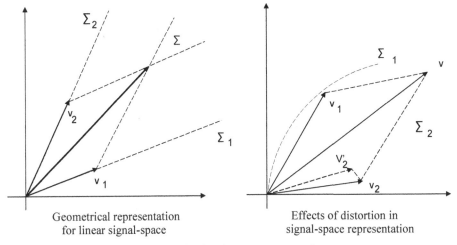

Geometrical representation	Effects of distortion in
for linear signal-space	signal-space representation

Fig. 2. Crosstalk effects of distortion in signal-space representation

The zero representation was used to study various forms of modulated waves. This representation was based on the principle of factorization in terms of the Fourier series expansion, and on the property of zero-pattern superposition that results from the product of two or more signals. Multiplicative processes are the most amenable to a zero-based description. Special attention was also paid in the study of crosstalk effects, Fig. 2 [4]. Tensor analysis was employed in the multi-space in which the zeroes of multiplicative signals are located. The evaluation of crosstalk in terms of tensor forms resulted in an advantageous simplification in the calculation procedure.

I consider that the use of CAPE in this case constituted real innovation in a field lacking mathematical models. I would even go as far as to say that this advance is particularly significant as it still applies today.

2.2 The Stochastic Computer

The *next example* illustrates CAPE's ability to introduce new concepts in computing techniques. As a departure from conventional digital or analog computing technologies, the stochastic (random-pulse) computer utilizes logical elements (gates) to process the analog magnitude that has been chosen to represent the variables (Fig. 3). The aforementioned analog magnitude is the probability of pulse-occurrence in a train of random pulses [5]. The variable value is recovered by averaging the stochastically coded variable over a period of time that is assumed to be stationary. It is readily apparent, for instance, that given that two statistically independent stationary random-pulse trains drive the two inputs of an AND gate, once the output pulse train is eventually reshaped it will have a probability of occurrence equal to the product of the probabilities of the incoming inputs [6]. A straightforward multiplier is thus obtained.

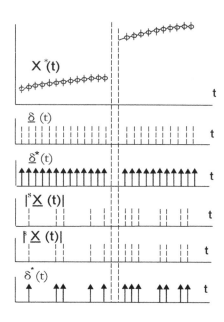

Fig. 3. Stochastic representation of variables

A random floating point stochastic coding was proposed through the use of weighted probabilities in order to increase the dynamic range of the variables to values of less than 1. The implementation of generalized (random floating point) stochastic coding made it possible to generate arithmetic operators (addition and subtraction) in a straightforward fashion, as well as to come up with the product and quotient. Integration and derivation can be easily performed by means of a bidirectional counter with weighed inputs followed by a digital to stochastic converter.

In order to generate functions in stochastic computation, a highly stimulating method that is *unique* to this technology was envisaged. In the case of linear stochastic conversion, the random-pulse train is a true stochastic representation of variables. However, this requires a uniform probability density function. This means that white Gaussian sequences, which were very difficult to obtain, must be used.

Alternatively, if at the stochastic conversion stage the cumulative probability function of a sampled random noise is not linear, but is instead an arbitrary monotone increasing function, the encoded variable would be the stochastic representation of this function, which is thus directly obtained. [7].

The problem of function generation is related to the problem of generating functional noises with specified cumulative probability functions. Pseudo-random *dinary* pulse-trains were proposed instead of random binary pulse-trains. This noise can be easily generated using the technique of maximum-length sequences, which is attained through the use of shift registers with appropriate feedback paths. The physical implementation was a module 127, three-stage shift register, which directly generates a maximum length sequence of 127^3-1 numbers of seven digits (equivalent to 127 levels) thus giving a computing accuracy of 0,1.

I consider multichannel stochastic computation to be *a wholly innovative concept* with practical evidence in control applications. Traffic control is a major application of this technique. Strangely enough, the whole traffic control project was licensed to the Russians. Ironically, it was to be implemented in Moscow in 1969, which at that time had a very low traffic density.

2.3 Modeling Macromolecules

My next example is closely connected with the extraordinary experience of working with two Nobel Prize winners, Maurice Wilkins and Jean Hanson, at King's College, London. Maurice Wilkins, a brilliant physicist, continued his excellent work on the *X-ray fiber diffraction of macromolecules*. In 1966, he and his PhD student, John Pardon, invented the toroidal camera, which made it possible to obtain precise, low-angle diffraction patterns that gave an accurate picture of long-pitch helical molecules such as nucleohistones. The remaining problem was to *find accurate structural models of complex macromolecules*, which led to a *refined tertiary structure of biopolymers*.

The advances in computer power and satellite technology were enthusiastically adopted by leading biophysicists around the world. I then worked with Wilkins at King's College and went on to work with Struther Arnott at Purdue University on DNA forms and complexes Fig. 4.

Fig. 4. B-DNA X-ray fiber diffraction pattern

A working model and strategy to overcome these issues were once again necessary. Arnott (an ex–fellow of King's College) and Peter Campbell came up with the most successful strategy for modeling and refining macromolecular structures [8]. As a result they created the core of the *linked-atom-least-squares* program (LALS), which was subsequently improved by several others, *myself included*.

The basic repeating structure – the nucleotide residue – was established by defining six conformational angles with which to build the anti-parallel chains and the glycosidic angle that linked them to the sugar rings. Five additional degrees of freedom made it possible to move the configuration of sugar to C'2 endo and C'3 endo puckering [9].

As part of my work involved in the *prediction of a rich variety of DNA configurations*, intensities along the different layers of this reciprocal space were mathematically calculated. This was done by taking the inverse of the Fourier transform, which was approximated by Bessel functions in the reciprocal space. As a final outcome, the radius and position, or phase, of each molecule's atom was obtained. The *model optimization* – the refined molecule structure – was obtained by minimizing the difference in structure amplitudes subject to the relative weight of the observations. Stereochemical acceptability was ensured by taking into account the interatomic distances d_j, which was calculated using a Buckingham energy function.

It was extremely rewarding to see that the *predicted variability did indeed correspond to physical structures* [10]. The two "classical" forms of DNA – A and B – gave rise to a variety of intermediate structures, which coincided with the actual binding of proteins (Table 1).

The comment "*a small step for mankind, but a giant step for Luis*" were dedicated to me by Struther Arnott when, by chance, I discovered *heteronomous DNA*, popularly called Z-DNA [11]. This discovery was related to the puckering of the sugar rings that changed from one nucleotide to the next. This caused real "kinks" in the structure that resulted in a superhelical structure, which had been theoretically predicted in the case of nucleohistones.

The Z-DNA fragments may constitute the building blocks that are embedded in the classical A and B forms of DNA, which give rise to hybrid structures. This *revolutionary* vision of DNA was reached rapidly, largely thanks to the advances made in computing power. Instead of the IBM 7094 with 170 kB that had been used in the past, the 8MB Cray supercomputer was developed. This development was superseded by the Cray Y-MP in 1988, which performed at a speed of 1Gflop. However, the enormous importance attached to this research subject also meant that the consolidation of CAPE tools relied heavily on contributions from other disciplines, such as quantum chemistry, which is the case today.

Table 1 DNA variety and variability

Family	Furanose Conformations	Conformational Genera	Number of Congeneric Species	Helical Characteristics	
				h(nm)	t(°)
A	C3' - endo	$t\,g^-\,g^-\,t\,g^+\,g^+\,a$	16	0.26-0.33	30.0-32.7
		$t\,g^-\,t\,t\,t\,g^+\,a$	1	0.31	36
B	C2' - endo	$t\,t\,g^-\,t\,g^+\,t\,a$	4	0.30-0.34	36.0-45.0
		$t\,t\,t\,t\,t\,t\,a$	1	0.33	48

The Z-DNA fragments may constitute the building blocks that are embedded in the classical A and B forms of DNA, which give rise to hybrid structures. This *revolutionary* vision of DNA was reached rapidly, largely thanks to the advances made in computing power. Instead of the IBM 7094 with 170kB that had been used in the past, the 8MB Cray supercomputer was developed. This development was superseded by the Cray Y-MP in 1988, which performed at a speed of 1Gflop. However, the enormous importance attached to this research subject also meant that the consolidation of CAPE tools relied heavily on contributions from other disciplines, such as quantum chemistry, which is the case today.

2.4 Modeling Fluidized-Bed Reactors

Modeling fluidized beds was complex enough, but coupled with the study of gasification reactions, the system was doubly complicated. Levenspiel (in 1976) encouraged me to *embark on a still more complex journey*: to trap the bubbles that deteriorated the fluidized bed's performance using a magnetic field [12].

A model already plagued with empirical correlations would not dare to reject meta-models that were based on laboratory experiments! Although the appearance of fluidization curves was similar for fluidization with and without a magnetic field, two important phenomena were observed. On the one hand, bench scale experimentation demonstrated that a "calming" zone could be reached within a certain range of magnetic field intensity. Within this range, the fluidized bed becomes stabilized and by-passing bubbles of the gas carrier vanish.

If the magnetic field was switched off, bubbling increased and turbulence was much greater. The gas limit velocity at which bed expansion was obtained without turbulence was named the "transition velocity u_b", after which bubbles would dilute (Fig. 5). The transition velocity could take values up to eight times the minimum fluidization velocity. It can be predicted as a function of the bed porosity, the angle of mean velocity u in channels and vertical axis, and the magnetic field intensity H [13].

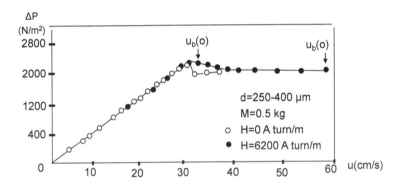

Fig. 5. Magnetically stabilized Fluidized bed behavior and transition velocity

A better knowledge of the structure of this type of fluidized bed (the stabilized bed) was achieved. The state of the bed could be described as falling between two limiting zones: the bed with particles situated at random with gas flowing though tortuous interstitial channels and the bed that forms ordered arrays of particles like chains with gas flowing straight through the rectilinear channels, (Fig. 6).

Applications were found in mixed systems containing magnetizable particles for the modeled, high-performance fluidized bed (Fig. 7). *A promising application was our incipient work on upgrading residual materials by thermal treatment in fluidized beds*. The mixtures consisted in refuse coal mixed with waste wood from different sources [14].

Sulfur abatement was efficiently achieved using a cheap catalyzer as an adsorbent, whose active component was iron oxide-based. The real problem that jeopardized the industrial application of this method was economy of scale. However, it gave us new insight into a complex system that is being incorporated into our current work to obtain clean hydrogen from waste materials. I consider this advance the result of the *evolution of several CAPE concepts and the hybridization of several techniques*.

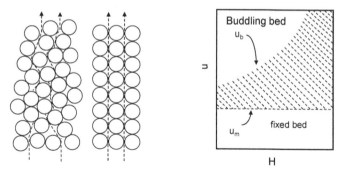

Fig. 6. Bed modeling: At left, spatial rearrangement of particles; at right, the three zone diagram

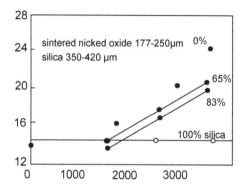

Fig. 7. Mixed particles systems applications

2.5 Modeling Batch processes

Back in the nineteen eighties, David Rippin whispered in my ear that "*the future is in batch*".

I will always be grateful to David for pointing me in the direction of a topic that at the time seemed an anachronism to engineers who were striving to achieve professional excellence by retrofitting batch designs into continuous operating processes. I was also very impressed by the amount of work already done by David, which he had written in German. He sent me his work in two big boxes, which I keep to this day. Most of this work had never been published, with the exception of that done on multi-batch and a few internal reports. This may also explain why we "started from the top", *as I was once told by Rex Reklaitis* at a meeting in Cambridge in 1988.

The truth of the matter is that after reading Rippin's material, I felt I had to start modeling the scheduling of a *multipurpose plant*. I was once again fortunate in that Manolo Lázaro was my first PhD student in the field of batch knowledge. He successfully identified the best production scheduling, out of the 451 feasible alternatives, for a real multi-purpose batch plant manufacturing three large volume products, being two of them dependent on production intermediates (Fig. 8) [15].

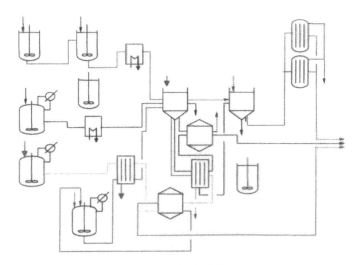

Fig. 8. Multipurpose batch plants planning and scheduling

David Rippin once called our work on batch modeling, and indeed, batch work in general, as *"filling in the holes"*. This was at ESCAPE 2 in Toulouse, in 1992. He was essentially right, since rigorous batch models were almost non-existent and the complexity of the problem did not make it possible to find solutions for specific models in reasonable computing times.

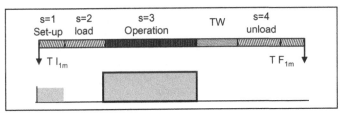

Fig. 9. Detailed scheduling model

For instance, it was not until 1993 that Moisès Graells [16] formulated an accurate representation of the subtasks examined in every batch process task (Fig.9). Obviously, the inclusion of this detailed representation of tasks in the mathematical production-scheduling model resulted in a highly complex formulation, and long computing times were necessary to solve it. Then a most complex **textile Company** manufacturing socks application came out. Specifically, the problem to solve includes planning for long- and short-term detailed scheduling for large parallel multiproduct facilities of a textile industry that manufactures 12,700 families of products, and whose main stage (called "Weaving") is constituted by 450 processing dedicated units working in parallel that require displacement and onsite installation for each family of products to achieve an optimal "chrono".

A Multi-objective set and the additional on-line information coming from actual plant operation allows any incident that occurs during the manufacturing to be adequately treated. Those two years (1992-93) *led to a breakthrough in the detailed modeling of a large-scale industrial application with satisfactory results* [17].

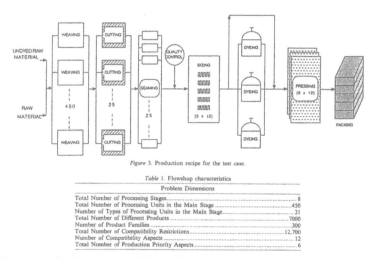

Figure 3. Production recipe for the test case.

Table 1. Flowshop characteristics

Problem Dimensions	
Total Number of Processing Stages	8
Total Number of Processing Units in the Main Stage	450
Number of Types of Processing Units in the Main Stage	21
Total Number of Different Products	7000
Number of Product Families	300
Total Number of Compatibility Restrictions	12,700
Number of Compatibility Aspects	12
Total Number of Production Priority Aspects	6

Fig. 10. Integrated planning and scheduling of a large industrial facility

Conclusions

I would like to end by saying that my instinct tells me that *breakthrough discoveries do not readily happen in engineering*, even more so in the case of Computer Aided Process Engineering (CAPE), as a second paper presented in CIMPS will corroborate. This is mainly *because new ideas* in this field *are continuously undergoing a process of dynamic development until they become sufficiently mature to be adopted by our industrial and social fabric, which is precisely the underlying essence of engineering.*

Acknowledgements

Financial support received from the Spanish "Ministerio de Economía y Competitividad" and the European Regional Development Fund, both funding the research Project ECOCIS (ref. DPI2013-48243-C2-1-R), and from the "Generalitat de Catalunya" (AGAUR 2014-SGR-1092-CEPEiMA)" is thankfully acknowledged

References

1. Puigjaner, L.: Historical tour of PSE: 1965 - Present. Plenary conference upon receiving the Long Term Achievements Award at the closing ceremony of the European Symposium on Computer Aided Process Engineering-18, Lyon, France on June 4 (2008)
2. Babbage, C.: On the economy of machinery and manufactures. - 4. ed. enlarged. - London: Charles Knight, 1835. - XXIV, 408 p.; 18 cm. - Fra p. XII e p. XIII: Preface to the fourth edition.
3. Puigjaner, L.: Analytical Signals and Zero-Locus in Multiplexing Systems. M.Sc. University of Houston, Texas (1969)
4. Puigjaner, L.: Computational Method of Crosstalk effects in Multichannel Systems of Aerospace Communications. In Automatic Control in Space (Ed. J.A. Aseltine), Inst. Society of America, Pensylvania, 3, 797-807 (1970)
5. Ferraté, G.A., Puigjaner, L., Agulló, J.: Técnicas de Cálculo Estocástico en la Investigación Bioquímica. Anales Real Soc. Esp. Fis. Quím., Anales de Química, 65, 1174 (1969)
6. Ferraté, G.A., Puigjaner, L. Agulló J.: Introducción to Multichannel Stochastic Computation and Control. In Proc. IV World IFAC Congress, (Ed. Naczelna Organiczna, Techniczna, Warsawa, Varsovia 63, 40-54 (1969)
7. Ferraté, G.A., Puigjaner, L., Agulló J.: Function Generation in Stochastic Conversion." In Proc. V World IFAC Congress, (Ed. J. Axelby), Pergamon Press, London, 2, 1-8 (1972)
8. Campbell-Smith, P.J., Arnott, S.: LALS: A linked-atom least-squares reciprocal space refinement system incorporating stereochemical restraints to supplement sparse diffraction data, Acta Crystallographica Section A Foundations of Crystallography, 34(1), 3-11 (1978)
9. Puigjaner, L., Subirana, J.A.: Low Angle X-Ray Scattering by Disordered and Partially Ordered HelicalSystems, J. Appl. Cryst., 7 (2), 169-173 (1974)
10. Subirana, J.A., Puigjaner, L.: Circular Superhelical DNA" Nature, 267, 727 (1977)
11. Arnott, S., Chandrasekaran, R., Hall, I.H., Puigjaner, L.: Heteronomous DNA, Nucleic Acids Res., 11 (12), 4141-4155 (1983)
12. Arnaldos, J., Casal, J., Lucas, A., Puigjaner, L.: Magnetically stabilized Fluidization: Modelling and Application to Mixtures, Powder Technology, 44, 57-62 (1985)
13. Lucas, A., Arnaldos, J., Casal, J., Puigjaner, L.: High Temperature Incipient Fluidization in Mono and Polydispare Systems, Chem. Eng. Commun. 41, 121-132 (1985)
14. Oliveres, M., Alonso, M., Recasens, F., Puigjaner, L.: Modeling and Simulation of the Styrene-Acrylonitrile Emulsion Polymerization Kinetics. The Chemical Engineering Journal, 34, 1-9 (1987)
15. Lázaro, M., Espuña, A., Puigjaner, L.: A comprehensive Approach to Multipurpose Batch Plants Production Planning, Computers and Chemical Engineering, 13, 1031-1047 (1989)
16. Graells, M., Cantón, J., Peschaud, B., Puigjaner, L.: General approach and tool for the scheduling of complex production systems. In European Symposium on Computer Aided Process Engineering, Computers and Chemical Engineering, 22, 395-402 (1998)
17. Espuña, A., Puigjaner, L.: Solving the Production Planning Problem for Parallel Multiproduct Plants, Chem. Eng. Res. Des., 67 (6), 589-592 (1989)

CAPE Role in Engineering Innovation: Part 2-The Coming Revolution

Luis Puigjaner[1*], Edrisi Muñoz[2], Elisabet Capón-García[3]

[1] Department of Chemical Engineering, Universitat Politècnica de Catalunya ETSEIB, Avda. Diagonal, 647, E-08028 - Barcelona, Spain
[2] Centro de Investigación en Matemáticas A.C., Jalisco S/N, Mineral y Valenciana 36240, Guanajuato, Mexico
[3] Department of Chemistry and Applied Biosciences, ETH Zurich, Wolfgang-Pauli-Str 10, 8093 Zurich, Switzerland
luis.puigjaner@upc.edu, emunoz@cimat.mx, elisabet.capon@chem.ethz.ch

Abstract. The advancement of science in the past century gave rise to a number of revolutionary discoveries that deeply affected the way of life of our society. A brief description of today's state of the art software in Process Systems Engineering: *meta-models* providing an integrated/unified decision support framework at all levels (strategic, tactic, operational) to the industry is presented in this second part. Additionally, knowledge as base of future developments and the efforts that CAPE approach has envisaged in order to develop systems thinking and systems problem solving are also introduced. Finally, it is also presented the research project called *Batch Process Ontology Framework based on ISA standards*, which has evolved along different fields of CAPE, regarding optimization methodologies, batch process, monitoring & control, scheduling, supply chain design, life cycle assessment, mathematical programming, modeling, and operations research.

Keywords: Software Engineering, Computer aided process engineering, Process systems engineering.

1 Introduction

At the end of the last century, methods, tools and software received soon a great impulse in the recently coined Computer Science. Subsequently, concepts and tools in computer aided design (CAD) and computer aided manufacturing (CAM) started to emerge [1]. These concepts and the applications that emerged as a result were labeled computer-aided process operations, computer integrated process engineering and a variety of additional names that highlighted the "use of computers". In 1996, the European Federation of Chemical Engineers created a Working Party (WP) on the topic "Computer Aided Process Engineering" (CAPE). Since 1996 international

© Springer International Publishing AG 2017
J. Mejia et al. (eds.), *Trends and Applications in Software Engineering*, Advances in Intelligent Systems and Computing 537, DOI 10.1007/978-3-319-48523-2_9

symposiums every year on CAPE related topics, computers are now routinely applied throughout the entire spectrum of process and product engineering activities covering chemical, petrochemical, bio-chemical, pharmaceutical industries (Fig. 1) and thereby, reflecting the success of the Working Party's activities[1].

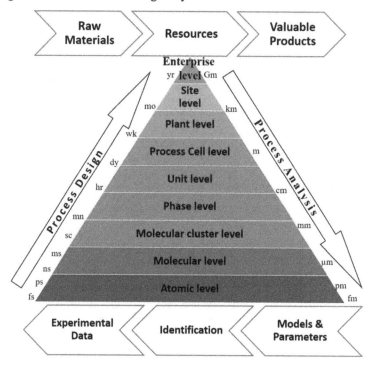

Fig. 1. Computer Aided Process Systems domain

Since its conception in 1996 the CAPE-WP has been intimately related with the concept of Process Systems Engineering (PSE). Takamatsu [2] defines PSE as an academic and technological field related to *methodologies* for chemical engineering *decisions.* Such methodologies should be responsible for indicating "how to plan, how to design, how to operate, how to control any kind of unit operation, chemical and other production process or chemical industry itself". Other authors have reached similar definitions, where planning, design, operation and control are key activities of these academic disciplines [3, 4, 5].

From a holistic point of view, it is the **system** under concern, which is at the root of both fields. In this sense, PSE approach pervades a wide spectra of science and engineering topics, expanding the horizon of human knowledge to chemical, energy and bioenergy, biology, telecommunications, mechanics, mining, pharmaceuticals etc. what will be described as follows [6].

Indeed, the use of computers greatly boosted all fields of knowledge. As noted in Part 1 of this work, Biology is perhaps the most outstanding case. In a relatively short

---

[1] See CAPE-WP web page at: http://www.cape-wp.eu/page.php?1

space of time, teaching and research in biology and the natural sciences, which had for many years been entrenched in very classical approaches, became the most avid consumers of computing power. Thus, after serving for 100 years as traditional teaching centers, as was the case of the Harvard Museum of Natural History, *museums have ceased to be houses of learning and research.*

However, this newly found resource *did not contribute to a radical innovation of applications.* After all, these applications were merely a vague reflection of the potential of computer sciences.

Engineering is *no exception.* At the beginning, engineering drew on knowledge from other fields, essentially mathematics, physics and chemistry. It has thus always been a very rich discipline as it takes great discoveries from other areas of knowledge and gives them added value by finding useful applications and making the appropriate changes of scale to them. Thus, engineering has also taken the same approach to computer science.

One may be aware that *radical innovation is not unique to engineering.* So, it comes as no surprise that the Nobel Prize has been awarded to mathematicians, physicists, chemists, biologists, social scientists and even political scientists, but as yet no engineer has won this prize.

However, CAPE *has done a great deal* enhancing the major role played by engineering *in the evolution* of new discoveries and *in their rapid adaptation* to the needs of mankind.

1. The Present

2.1. Modeling the Supply Chain

Current trends in CAPE's modeling approaches extends batch modeling, so that it reflects reality as closely as possible, continues today. The continuous progress made in modeling has brought about novel approaches to batch modeling. These new approaches include the integration of energy and the minimization of waste, in addition to the combination of batch control, integrated scheduling and control, integrated production scheduling and financial control, proactive and reactive online scheduling strategies, energy, water and waste minimization, and so on [7, 8, 9, 10, 11, 12].

It is now proving to be quite interesting to review the role of *transfer times*, which are usually ignored to simplify model representation. After exhaustive reflection on this matter, it has revealed itself to be critical in finding rigorous optimal solutions as it appears in recent publications [13, 14].

Next, we examine the supply chain. The role of the supply chain is recognized worldwide, and has been recently included as one of the key components of the AIChE Sustainability Index, according to a 2007 issue of CEP [15] (Fig. 2).

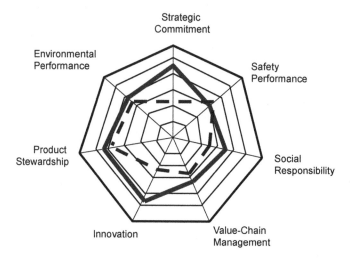

Fig. 2. Sustainability Global Index (AIChE): Brown color continuous line taken from Global Fortune 500 Chemical companies; blue broken line from U.S. Chemical manufacturing sector [15]

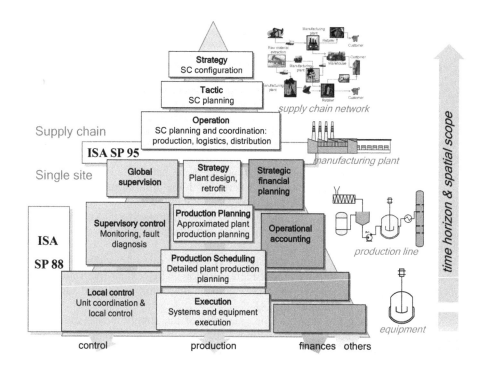

Fig. 3. Integrated Supply Chain with detailed components

In fact, the supply chain in its widest sense is the missing link between product and process engineering. This may become more apparent by looking at the one-dimensional pyramid described here (Fig. 3). The supply chain embraces production control, financing, the environment, negotiation, social aspects, product stewardship, innovation, safety performance and strategic commitment [16, 17, 18, 19]. What is more, it offers, or at least should offer, *a comprehensive view of the global scenario*. It thus avoids us having a fragmented perception of the overall problem, which results in *uncoordinated and even contradictory decisions*. Thus, we are finding real solutions for businesses that require *highly dynamic responses at all levels in terms of transparency and reliability*. This may seem to be a futuristic view that belongs in the realms of science fiction. However, looking back, one could have felt exactly the same way in attempting to solve the integrated long- and short-term production planning of a 7000-products manufacturing facility seen in Part 1.

Building blocks and supporting architectures have already started to emerge. They are also open to the incorporation of new components in the future. Moreover, a project has been proposed that is based on this big picture. In my view, *this serves as an inspiring example of the intrinsic, evolutionary nature of engineering*, which may give rise to *revolutionary discoveries* in this field.

3.   The Future

There are emerging fields in CAPE that are considered as main future key fields of relevance. At the same time, these key fields must tackle fundamental issues, such as computing infrastructure and industrial technology transfer. Some of these main fields of relevance include [20, 21, 22, 23, 24, 25]:

i)   *Modeling, Simulation and Optimization*, which drives to a more integration of process, equipment and product design framework. The main requirement of this field lies in the development of unifying modeling approaches, encompassing different time and space scales (ranging from micro-scale to macro-scale) by using multi-scale models incorporated in computer-aided modeling tools.

ii)   *Sustainable process synthesis*, supporting the generation and evaluation of different alternative process structures based on engineering experience, the support of multi-objective decision-making integrating areas of social, economic, environmental and processing, which are heavily influenced by risk and uncertainty.

iii)   *Process operations and management*, facing the agile management of process plants and the global supply chains, comprising a number of enterprises in different geopolitical situations. Besides, they must work on an integrated atmosphere, vertically across the automation hierarchy of a single process plant and horizontally along the supply chain. The expected result is an integrated designing process and its associated operational support system, which includes control, optimization and scheduling functionalities, data, information and knowledge.

iv)   *Information technology (IT)*, supporting engineering design and development processes involves big opportunities and potential for cost reduction and quality improvement in industrial design processes. A new generation of cost-effective and tailor-made supporting software solutions is suggested which reflect the culture and the specific work processes of an enterprise.

v)   *Numerical algorithms and computing paradigms* by suitable modular methods, which even take advantage of distributed and parallel computing architectures. Such a modular strategy would also support the use of multiple numerical methods tailored to the requirements of a partial model, comprising special structure of a selection of algebraic, differential, partial differential or integer-differential equations.

3.1 Knowledge as base of future developments

One of the CAPE´s main commitments to ensure a sustainable success of the above-mentioned fields is to develop *systems thinking* [26]. It refers to the development and exploitation of the global system based on tangible and intangible knowledge resources. Semantic technologies seem to offer an attractive platform for knowledge capturing, information management and work process guidance in the design processes including their associated control and operating support systems. They also support a smooth integration of information modeling and mathematical modeling in a single modeling framework. Those meta-models arise, which are focused in the analysis, construction and development of the frames, rules, constraints, models and theories applicable and useful for the modeling in a predefined class of problems.

Nowadays, there is a plenty number of commercial systems providing features such as process design, simulation, monitoring, execution, control and optimization among others, presented for different purposes and detail. The evolution of management system includes the integration of implementation methodologies and tools which must address the required capabilities for both, including automation, performance, and flexibility for the process side, and collaboration, query and retrieval, and taxonomy for the knowledge management side.

Important efforts in CAPE have been made in order to develop systems thinking and systems problem solving, but still have to be prioritized, rather than being the mere application of computational problem solving methods. There are many emerging areas where systems thinking and systems engineering methods and tools are most likely a key to success.

Particularly, CEPIMA group since few years ago has been working in this line. Since 2007 Edrisi Muñoz began the research project called *Batch Process Ontology Framework based on ISA standards*, which has evolved along different fields of CAPE, regarding optimization methodologies, batch process, monitoring & control, scheduling, supply chain design, life cycle assessment, mathematical programming, modeling, and operations research [27, 28, 29]. The aim of this work is the development of powerful decision support tools for the process industry. As a decision support tool, it must be capable of becoming a robust model, which interacts among the different decision hierarchical levels (strategic, tactic and operational) providing a unified framework of data and information levels integration. Thus, a user interface interacts with an ontological model (batch processes domain) on the client side via a listener pattern. When the model needs to be updated with new data, a Remote Procedure Call (RPC) client module invokes a request to the RPC server module. An interaction with the ontology and the collaboration API is made, in order to provide the requested data. The ontology model allows the relation among different actors of the framework. In addition, this framework enables the reuse of existing data for crating or adding new structure or single data (Fig. 4).

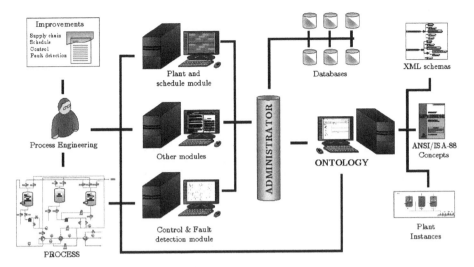

Fig. 4. First Ontology Model framework approach for PROCEL pilot plant (CEPIMA-UPC)

As ongoing results, the integration of various decision levels within a common representation facilitates the creation, storage and sharing of knowledge in a specific domain, and allows to improve the effectiveness of decision support systems. Thus, *the gap between analytical techniques* (e.g. optimization approaches) *and transactional systems* (e.g. ask processing and communicating data) *has been reduced.*

Furthermore, since the proposed enterprise ontology framework has been created using industrial standards, it has resulted in a well-behaved performance at capturing common understanding in conceptual design and at helping to utilize the relations that are mined from databases. Such properties help users to obtain data and information in a proper, fast and standard way.

This work offers the required effective techniques and technology for rapid integration of communication and management of multiple systems. Finally, the framework has been applied, in addition to process systems engineering, to different production areas, such as mining, chemical, automotive, pharmaceutical and software engineering, for the achievement of different specific tasks and objectives.

Conclusions

It is ironic to realize that Alfred Nobel, who established the Nobel Prizes, was an engineer, more precisely a chemical engineer that invented dynamite and other explosives. Notwithstanding, engineers have never been selected for the Nobel Award, while any other branch of sciences have been honored by it. Moreover, Nobel Award has been extended to social, artistic and other human activities. This relevant fact corroborates the conclusion of Part 1, where was stated that *breakthrough discoveries do not readily happen in engineering*, even more so in the case of CAPE. This is mainly *because innovation* in this field *is continuously undergoing a process of dynamic development until it proves to be sufficiently mature to be adopted by our industrial and social fabric, which is precisely the underlying essence of engineering,* the branch of sciences that has contributed the utmost to make *human work more human.*

Acknowledgements

Financial support received from the Spanish "Ministerio de Economía y Competitividad" and the European Regional Development Fund, both funding the research Project ECOCIS (ref. DPI2013-48243-C2-1-R), and from the "Generalitat de Catalunya" (AGAUR 2014-SGR-1092-CEPEiMA)" is thankfully acknowledged

References

1. Valentino, J.V., Goldenberg, .: Introduction to Computer Numerical Control 3/E. Prentice Hall, Upper Saddle River, New Jersey (1991)
1. Takamatsu, T.: The nature and role or process systems engineering. Computers and Chemical Engineering 7, 203–218 (1983)
3. Sargent, R.W.H.: What is chemical engineering? CAST Newsletter, 14(1), 9–11 (1991)
4. Grossmann, I. E., Westerberg, A.W.: Research challenges in process systems engineering. AIChE J., 46, 1700–1703 (2000)
5. Klatt, K.U., Marquardt, W.: Perspectives for process systems engineering—Personal views from academia and industry, Computers and Chemical Engineering, 33, 536–550 (2009)
6. Stephanopoulos, G., Reklaitis, G.V.: Process systems engineering: From Solvay to modern bio- and nanotechnology: A history of development, successes and prospects for the future. Chemical Engineering Science, 66 (19), 4272–4306 (2011)
7. Corominas, J., Espuña, A., Puigjaner, L.: A new look to Energy Integration in Multproduct Batch Processes", Computers and Chemical Engineering, 17S, 15-20 (1993)
8. Puigjaner, L., Espuña, A., Graells, M., Corominas, J.: Combined Energy and Waste Minimization in Multiproduct Batch Processes, Hungarian Journal of Industrial Chemistry, 21, 251-264 (1993)
9. Corominas J., Espuña, A., Puigjaner, L.: Method to Incorporate Energy Integration Considerations in Multiproduct Batch Processes, Computers and Chemical Engineering, 18, (1994)
10. Puigjaner L., Huercio, A., Espuña A.: Batch Production Control in a Computer Integrated Manufacturing Environment, Journal of Process Control, 4, 281-290 (1994)
11. Grau, R., Espuña, A. and L. Puigjaner.: Focusing in by-product recovery and waste minimization in batch production scheduling. Computers and Chemical Engineering, 18S, S271-S275 (1994)
12. Huercio, A., Espuña, A., Puigjaner, L.: Incorporating On-line Scheduling strategies in integrated Batch Production Control. Computers and Chemical Engineering, 19S, 609-614, (1995)
13. Grau, R., Espuña, A., Puigjaner, L.: Completion Times in Multipurpose Batch Plants Set-Up, Transfer and Clean-Up Times,. Computers and Chemical Engineering, 20S(5), S1143-S1148 (1996)
14. Ferrer-Nadal, S., Capón-García, E., Méndez, C., Puigjaaner, L.: Material transfer operations in batch scheduling. A critical modeling Issue, Ind. & Eng. Chem. Res., 47(20), 7721-7732 (2008)

15. Cob, C., Schuster, D., Beloff, B., Tanzil, D.: Benchmarking Sustainability. Chemical Engineering Progress, American Institute of Chemical Engineers, 38-42, (2007)
16. Guillén, G., Mele, F.D., Bagajewicz M., Espuña, A., Puigjaner, L.: Multiobjective supply chain design under uncertainty, Chemical Engineering Science, 60, 1535-1553 (2005)
17. Guillén, G., Pina, C., Espuña, A., Puigjaner, L.: Optimal offer proposal policy in an integrated suply chain management environment, Ind. & Eng. Chem Res., 44, 7405-7419 (2005)
18. Guillén, G., Badell, M., Espuña, A., Puigjaner, L. Simultaneous optimization of process operations and financial decisions to enhance the integrated planning/scheduling of chemical supply chains, Computers and Chemical Engineering, 30, 421-436 (2006)
19. Mele, F.D., Espuña, A., Puigjaner, L. Supply chain Management through dynamic model parameters optimization, Ind. & Eng. Chem. Res., 45, 1708-1721 (2006)
20. Mele, F.D., Guillén, G., Espuña, A., Puigjaner, L.: A simulation-based optimization framework for parameter optimization of supply-chain networks, Ind. & Eng. Chem. Res., 45, 3133-3148 (2006)
21. Guillén, G., Mele, F.D., Espuña, A., Puigjaner, L.: Addressing the design of chemical supply chains under demand uncertainty, Ind. & Eng. Chem. Res, 45 (22), 7566-7581 (2006)
22. Bojarski A., Laínez J., Espuña A., Puigjaner L.: Incorporating environmental impacts and regulations in a holistic supply chains modelling: An LCA approach. Computers & Chemical Engineering, 33 (10), 1747-1759 (2009)
23. Laínez J.M, Puigjaner L., Reklaitis G:V:, "Financial and financial engineering considerations in supply chain and product development pipeline management" Computers & Chemical Engineering 33 (12) ISSN: 0098-1354, pp.1999-2011 (2009)
24. Puigjaner, L., Laínez, J.M, Rodrigo, C.: Tracking the Dynamics of the Supply Chain for Enhanced Production Sustainability. Ind. & Eng. Chem. Res. 48 (21), 9556-9570 (2009)
25. Kopanos, G., Méndez, C., Puigjaner L.: MIP-based decomposition strategies for large-scale scheduling problems in multiproduct multistage batch plants: A benchmark scheduling problem of the pharmaceutical industry. European Journal of Operational Research 207 (2), 644-655 (2010)
26. Records, L. R.: The fusion of process and knowledge management the fusion of process and knowledge management, Business Process Trends, (2005)
27. Muñoz, E., Espuña, A., Puigjaner, L.: Towards an ontological infrastructure for chemical batch process management, Computers and Chemical Engineering, 34 (5), 668-682 (2010)
28. Muñoz, E., Capón-García, E., Laínez-Aguirre, José M., Espuña, A., Puigjaner, L.: Using mathematical knowledge management to support integrated decision-making in the enterprise, Computers and Chemical Engineering, 66, 139–150 (2014)
29. Muñoz, E., Capón-García, E., Laínez, J.M., Espuña, A. Puigjaner, L.: Supply chain planning and scheduling integration using Lagrangean decomposition in a knowledge management environment, Computers and Chemical Engineering, 72 (1), 52 – 67 (2015)

Using Design Patterns to Solve Newton-type Methods

Ricardo Serrato Barrera[1], Gustavo Rodríguez Gómez[2], Saúl Eduardo Pomares Hernández[2], Julio César Pérez Sansalvador[3], and Leticia Flores Pulido[4],

[1] Estratei Sistemas de Información, S.A. de C.V., Virrey de Mendoza 605-B, Col. Las fuentes, 59699, Zamora, Michoacán, México
[2] Instituto Nacional de Astrofísica, Óptica y Electrónica, Coordinación de Ciencias Computacionales, Luis Enrique Erro 1, 72840, Tonantzintla, Puebla, México
[3] Instituto Nacional de Astrofísica, Óptica y Electrónica, Laboratorio de Visión por Computadora, Luis Enrique Erro 1, 72840, Tonantzintla, Puebla, México
[4] Universidad Autónoma de Tlaxcala, Facultad de Ciencias Básicas, Ingeniería y Tecnología, Calzada Apizaquito, Colonia Apizaquito, 90300, Apizaco, Tlaxcala, México
{rsbserrato, grodrig, spomares, tachidok}@ccc.inaoep.mx, leticia.florespo@udlap.mx

Abstract. We present the development of a software system for Newton-type methods via the identification and application of software design patterns. We measured the quality of our developed system and found that it is flexible, easy to use and extend due to the application of design patterns. Our newly developed system is flexible enough to be used by the numerical analyst interested in the creation of new Newton-type methods, or the engineer that applies different Newton--type strategies in his software solutions.

Keywords: Newton-type methods, design patterns, object-oriented, software design, scientific software.

1 Introduction

Newton-type methods are a family of numerical methods widely used to solve nonlinear systems of equations, unconstrained optimisation and data fitting problems. The popularity of these methods has led to the development of software packages implementing variations of them, e.g., HOMPACK [1], TENSOLVE [2] and NITSOL [3] are software packages to solve optimisation and nonlinear problems, all of them implemented in Fortran using *the procedural programming paradigm*. This approach generates complex and hard to reuse interfaces, inappropriate data structures and code that does not captures the implemented algorithms [4]. Software packages such as COOL [5], PETSC [6] and OPT++ [4] are object-oriented implementations of Newton-type methods. When the object-oriented approach is applied focusing on the details rather than on the bigger picture the produced software is commonly highly coupled, hard to modify and difficult to understand.

Software design patterns are expert solutions to software design problems. The development of scientific software lead by the application of design patterns is not common due to the difficulty on the identification and mapping of problem-specific concepts into software patterns. The relations established by the objects identified at the design stage establish the structure and quality of the final software design.

© Springer International Publishing AG 2017
J. Mejia et al. (eds.), *Trends and Applications in Software Engineering*, Advances in Intelligent Systems and Computing 537, DOI 10.1007/978-3-319-48523-2_10

Software patterns guide us through the design stage, they expose relations between the entities that describe a problem in a particular context [7].

Currently, few works have introduced design patterns in the field of scientific software. Some have identified and proposed new patterns for the development of scientific software, [8]-[10]. Others have successfully applied design patterns for the development of scientific software, [11]-[13].

In this work we take the ideas and principles described by design patterns and apply them for the development of scientific software, we design a novel pattern object-oriented software system design for Newton-type methods. The key requirement is the flexibility of this design to permit the inclusion of new Newton-type methods. We use Martin's metric [14] to evaluate the abstractness (or generality) and the instability (or ability to reuse its existing parts) of the software system.

The remaining of this document is organised as follows: in section 2 we present the mathematical background regarding Newton-type methods and the variants considered in this work; in section 3 we present the development of the software system and the application of design patterns to solve software design problems; in section 4 we validate the newly developed software system; and in section 5 we present our main conclusions.

2 Mathematical Background

Newton's method may be considered as the standard technique used by the scientific, engineering and software development community to solve problems presenting nonlinear behaviour. We focus on three classes of nonlinear problems:

- Nonlinear equations problems involve to find x_* such that the vector-valued function F of n variables satisfies $F(x_*) = 0$.

- Unconstrained optimisation problems comprise to find x_* such that the real-valued function f of n variables satisfies $f(x_*) \leq f(x)$ for all x close to x_*.

- Nonlinear least-square problems require to find x_* such that $\sum_{i=1}^{m}(r_i(x))^2$ is minimised, r_i denotes the i-th component function of

$$G(x) = (r_1(x), r_2(x), \ldots r_m(x)), x \in R^n, m \geq n.$$

In Algorithm 1 we present the generic form of Newton's method given in [15].

Algorithm 1. Newton's method generic steps.

Require. Initial guess x_0.

1: Initialise iteration counter $k = 0$.

2: **while** stopping condition is not satisfied **do**

3:    Compute Newton direction s.

4:    Calculate the step length λ.

5:    Get a new approximation $x_{k+1} = x_k + \lambda s$.

6:    Increase the iteration counter $k = k + 1$.

7: **end while**

8: Return x_k as the approximated solution of x.

The selection of particular strategies, such as line search or trust regions methods, to compute the Newton direction (step 3) and the step length (step 4) gives rise to specific variations of Newton-type methods; each solving an specific nonlinear problem.

In what follows we use $JF(x) = (\partial f_i / \partial x_j)_{i,j}$ to represent the Jacobian matrix of the function $F(x)$. The gradient of a function $f(x)$ is denoted by $\nabla f(x) = (\partial f / \partial x_1, ..., \partial f / \partial x_n)^T$, and the Hessian of $f(x)$ is the matrix $Hf(x) = (\partial^2 f(x) / \partial x_i \partial x_j)_{i,j}$.

3 Newton-type Methods Software Design

We start by decomposing the problem into sub-systems by applying the commonality and variability analysis (CVA) of Coplien [16] to identify key concepts and study their variations in different scenarios of the problem domain. Then we apply the Analysis Matrix of Shalloway [7] to determine relationships between the previously identified concepts. We study these relations to gain an indication of potential design problems and identify design patterns that may be applied to solve them. The relations found at this stage dictate the structure of the software system design.

3.1 Base System Design

The base system design is generated from the sub-systems decomposition of the generic form of Newton's method, see Algorithm 1. The studied variations of Newton-type methods have steps 3, 4 and 5 in common, what varies is their particular implementation. By applying a CVA we identify key concepts and their variations in different scenarios, see Table 1.

Table 1. CVA for Newton methods.

Scenarios	Concept: Newton direction s	Concept: Step length λ
Line search methods	N/A	Compute a step length to guarantee convergence.
Trust region methods	The Newton direction is obtained by solving the linear or quadratic model within a trust ratio.	Constrain the step length by the trust radio.
Damped methods	Solve the system $JF_k \Delta x_k = -F_k$ and find the Newton direction, where $\Delta x_k = x_{k+1} - x_k$.	Use line search methods to obtain the step length.
Quasi-Newton methods	Obtain the Newton direction by solving the system $A\Delta x_k = -F_k$, where A is a matrix representing the Jacobian or the Hessian matrix of F_k.	Obtain the step length using search or trust region methods.
Inexact methods	The Newton direction is obtained by solving the system $JF_k \Delta x_k = -F_k$ using an iterative method.	Use a line search method to obtain the step length.

Based on Algorithm 1 we identified two additional concepts from those presented in Table 1: the stopping condition and the evaluation function, see Figure 1.

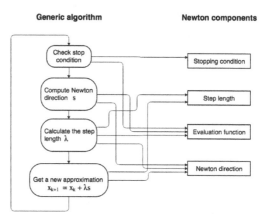

Figure. 1. Relations between the identified concepts or *Newton components* and the generic steps of Newton's method.

Each identified Newton component represents a sub-system. We observe that the Newton-type methods studied in this work share the same generic structure; an specific Newton-type versions can be created by varying the Newton components in the generic steps of Newton's method. From this statement we recognise the *template method* pattern [17] to represent the generic structure of Newton's method, and the *facade* design pattern [17] to define a simple and general interface for each of the steps or Newton components of the generic Newton algorithm steps.

Now suppose that we have two different implementations for the same Newton component, one used for development and another used for high-performance applications. In order to provide a *black-box* design where we hide the technical details of the implementation to the users we apply the *bridge* design pattern which decouples the abstraction from the implementation and allows them to vary independently, [14]. The base software system design for Newton-type methods is presented in Figure 2.

3.2 Newton's Method Sub-systems Software Design

Nonlinear methods. We observe that the three nonlinear problems presented in section 2 are particular or general cases of each other; *e.g.*, a nonlinear least-square problem is a particular case of an unconstrained optimisation problem. Consider a function $F(x) = 0$ and define $f = 1/2\|F\|^2$, finding an x such that $F(x_*) = 0$ is equivalent to find an x_* such that $f(x_*) = 0$. We represent this relation as a *transition* between different nonlinear problems; in this case from a nonlinear least-square problem to an unconstrained optimisation problem, see Figure 3.

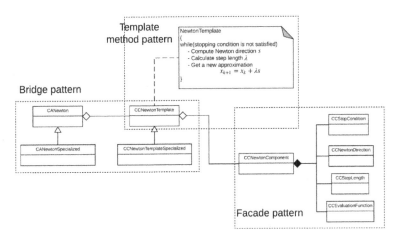

Figure. 2. Base system design for Newton-type methods showing the application of the template, facade and bridge design patterns.[1]

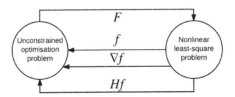

Figure. 3. Transitions between a nonlinear least-square problem and an unconstrained optimisation problem. The nonlinear functions and its derivatives are handled accordingly to the nonlinear problem to solve.

We recognise that three studied nonlinear problems are mathematically equivalent, thus they can be reduced to the solution of an unconstrained optimisation problem. However, following this approach requires the user to perform this transformation. Our goal is to develop a software system that provides the user with the tools to treat and solve his nonlinear problem using different strategies (without him having to implement these strategies), thus he can select the one that satisfies its application requirements.

In order to identify key concepts and relations between the three studied nonlinear problems we performed a CVA, see Table 2. We observe that for a particular scenario we compute either the Jacobian, the gradient or the Hessian of the nonlinear function. We recognise and apply the *state* design pattern [17] to implement transitions between strategies to handle the nonlinear function and its derivatives. Additionally, the user may provide the analytical form of the derivative or we could approximate it via finite differences or quasi-Newton method updates, we added these strategies to the software system design defined by the state pattern, see Figure 4.

---

[1] We use the prefix CA and CC to indicate abstract and concrete classes, respectively.

Table 2. CVA for nonlinear problems.

Scenarios	Concept: Nonlinear equations problem	Concept: Unconstrained optimisation problem	Concept: Nonlinear least-square problem
Function	$F:R^n \to R^n$	$f:R^n \to R$	$F:R^n \to R^m, n<m, f=\frac{1}{2}\|F\|^2$
First derivative	$JF(x)$	$\nabla f(x)$	$\nabla f(x) = JF(x)^T F$
Second derivative	N/A	$Hf(x)$	$Hf(x) = JF(x)^T JF(x) + \sum_{i=1}^{m} f_i Hf_i(x)$

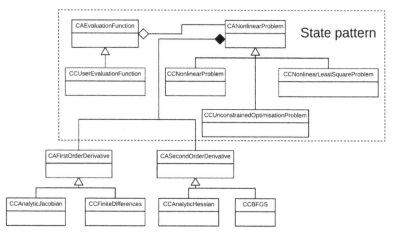

Figure. 4. Software system design implementing the state pattern to allow transitions between the studied nonlinear problems.

The software system design developed in this section correspond to the Newton component: evaluation function, presented in Figure 1.

Line Search Methods. Line search methods are strategies to find the step length λ to move along a direction s_k. The common schemes are based on bisection and interpolation. In order to determine whether the selected step length is appropriate, Wolfe, curvature and Goldstein test are applied [18], [19]. In Table 3 we present the concepts and its variations for the line search method scenario.

The *step length test condition* is applied as part of the computation of the step length, however, the test condition can vary independently of the step length approximation method. The *strategy* design pattern [17] allows us to define a family of algorithms, encapsulate them and make them interchangeable. We use a *double-strategy*, one strategy to encapsulate the step length approximation methods, and another strategy to encapsulate the decreasing condition methods, see Figure 5.

Table 3. CVA for line search methods.

Scenario	Concept: Step length approximation	Concept: Step length decreasing condition
Line search methods	Bisection Quadratic interpolation Cubic interpolation	Wolfe Goldstein Curvature condition

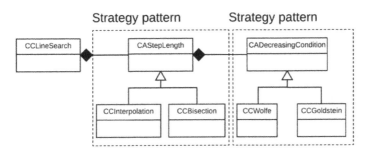

Figure. 5. A double-strategy pattern for the implementation of the selection of the step length.

The resulting software system design from this section corresponds to the Newton component: step length, presented in Figure 1.

Trust region Methods. Trust region methods are based on constructing a model to approximate a function $f(x)$ in a region around x_k. These methods can be reduced to solve a constrained minimisation problem

$$\min_{s \in R^n} m(s) = f(x_k) + \nabla f(x_k)^T s + \frac{1}{2} s^T H f(x_k) s \tag{1}$$

such that $\| s \| < \Delta$, where $\Delta > 0$ is the trust region radius. In Table 4 we present the identified concepts and its variations associated with the trust region scenario.

Table 4. CVA for trust region methods.

Scenario	Concept: Solve constraint problem	Concept: Update trust region radius
Trust region methods	Cauchy point method *Dogleg* methods Two-dimensional sub-space minimisation methods	Adaptive methods using threshold [18]

We encapsulate the methods that solve the minimisation problem and provide them with the same interface to make them interchangeable. We apply the *strategy* pattern to handle the methods to update the trust region radius and facilitate the addition of future methods. We apply the *adapter* design pattern to reuse the methods to solve nonlinear unconstrained optimisation problems from the nonlinear methods section; this pattern adapts the interface of an object such that it can be used in different contexts, [17], see Figure 6.

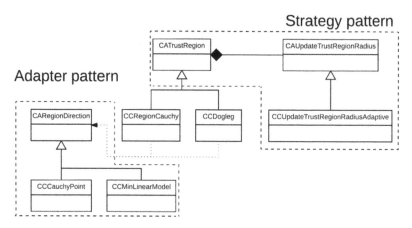

Figure. 6. Trust regions methods implemented via the strategy and the adapter pattern.

The resulting software system design from this section corresponds to the Newton component: Newton direction, presented in Figure 1.

Instantiating Objects and Interaction with External Packages. We use three more design patterns: the *abstract factory* to facilitate the creation and configuration of objects; the *singleton*, to supply a unique access point to all the factories; and the *adapter* pattern, to support the interaction with external packages for high-performance computations. The details of the application of these patterns are presented in [20] (the master's degree thesis of the first author).

4   Evaluation of the Software System Design

The developed software system design is formed by seventeen packages, see details in [20]. Martin defines in [14] the level of abstractness of a software system as

$$A = \frac{N_a}{N_c} \tag{2}$$

where N_a is the number of abstract classes in the package and N_c is the total number of classes in the package. The range of A is [0, 1]. $A = 0$ indicates a concrete package and $A = 1$ an abstract package. This metric is related to the capacity of extension of a software system, the more abstract the package the easier to extend. We also measure the instability of each package which is defined by Martin [14] as

$$I = \frac{C_e}{C_e + C_a} \tag{3}$$

where C_e is the number of classes inside the package that depend on classes outside the package, and C_a is the number of classes outside the package that depend on classes inside the package. The range of I is [0, 1]. $I = 0$ indicates an stable package and $I = 1$ an unstable package. This metric shows the ability of a package to support change. Martin combines these two metrics as $D = |A + I - 1|$, where $D = 0$ indicates a package easy to adapt or extend, and $D = 1$ a package difficult to adapt or modify, see Table 5.

Table 5. Martin's metric applied to the main packages of the architecture.

Package name	A	I	D
TrustRegionMethods	0.29	0.80	0.09
BaseArchitecture	0.75	0.31	0.06
LineSearchMethods	0.20	0.70	0.10
NonlinearMethods	0.29	0.46	0.25

We observe that most of the packages are near the main sequence, in particular, the package BaseArchitecture has $D = 0.06$, which indicates that the main package of the system is easy to extend, reusable and does not overuse abstraction.

5 Conclusions

We have presented the development of a software design for Newton-type methods; we applied eight design patterns from the book of Gamma *et. al.* [17]. The *template method pattern* defines the generic structure of the three studies Newton-type methods, the *facade* pattern supplies a simple interface for the Newton components, the *bridge* pattern allow us to implement different versions of a method to target the interest of different users, the *state* pattern hides the details of computing the first and second order derivatives of nonlinear functions, the *strategy* pattern allows us to change algorithms to compute the step length and decreasing condition in line search methods, it also let us add new methods to update the trust region radius, the *adapter* pattern provides a medium to communicate with third-party software libraries, it also allows us to reuse the strategies to compute the Newton direction from trust region methods, the *abstract factory* pattern provides an interface to create and configure the objects of the software system, and finally, the *singleton* pattern provides a unified and single interface for the communication with the factories.

A main contribution of this work is the identification and application of the *state* pattern for the development of scientific software, to the best of these authors knowledge the identified instance of this pattern has not been reported in related works. The instability and abstractness values of this pattern are those of the *NonlinearMethods* package, the one implementing the *state* pattern. The results show that the system design is stable enough to be extended without loss of flexibility. With the design of the presented software system we demonstrate that the knowledge of the scientific expert can be exploited by the software engineer through the application of design patterns to generate simple, flexible and effective object-oriented software. As part of our future work is the application of parallel technologies, integration of third-party state-of-the-art software libraries, use of templates and code optimisation techniques for the development of high-performance numerical software.

Acknowledgments. We sincerely thank the observations of the two anonymous reviewers that helped to improve and clarify this work.

References

1. L.T. Watson, S.C. Billups and A.P. Morgan, Algorithm 652: hom-pack: a suite of codes for globally convergent homotopy algorithm, ACM Trans. Math. Soft., vol. 13, no. 3, pp. 281-310 (1987).
2. A. Bouaricha and R.B. Shnabel, Algorithm 768: tensolve: a software package for solving systems of nonlinear equations and nonlinear least-square problems using tensor methods, ACM Trans. Math. Soft., vol. 23, no. 2, pp. 174-195 (1997).
3. M. Pernice and H. F. Walker, Nitsol: a Newton iterative solver for nonlinear systems, SIAM J. Sci. Comp., vol. 19, no. 1, p. 302 (1998).
4. J. Meza, R. Oliva, P. Hough and P. Williams, Opt++: an object-oriented toolkit for nonlinear optimization, ACM Trans. Math. Soft., vol. 33, no. 2, pp. 12-27 (2007).
5. L. Deng, W. Gouveia and J. Scales, The cwp object-oriented optimization library, Center for Wave Phenomena, Technical report (1994).
6. S. Balay, W. Gropp, L. McInnes and B. Smith, Petsc 2.0 users manual, Argonne National Laboratory, Technical report (1995).
7. A. Shalloway and J. Trott, Design Patterns Explained: A New Perspective on Object-Oriented Design (2Nd Edition) (Software Patterns Series). Addison-Wesley (2002).
8. C. Blilie, Patterns in scientific software: an introduction, Compt. Sci. Eng., vol. 4. no. 3, pp. 48-53, 2002.
9. G. Rodríguez-Gómez, J. Muños-Arteaga and B. Fernández, Scientific software design through scientific computing patterns, in Fourth IASTED International Conference, Hawai, USA, 2004.
10. T. Cickovski, T. Matthey and J. Izaguirre, Design patterns for generic object-oriented scientific software, Department of Computer Science and Engineering, University of Notre Dame, Technical report, TR05-12 (2005).
11. V. K. Decyk and H. J. Gardner, Object-oriented design patterns in Fortran 90/95, Comput. Phys. Commun., vol. 178, no. 8, pp. 611-620 (2008).
12. J. Pérez-Sansalvador, G. Rodríguez-Gómez and S. Pomares-Hernández, Pattern object-oriented architecture for multirate integration methods. In CONIELECOMP, Puebla, Mexico, 2011, pp., 158-163.
13. D. Rouson, J. Xia and X. Xu, Scientific Software Design, 1st. New York, USA: Cambridge University Press, 2011.
14. R. Martin, Agile software development: principles, patterns, and practices. NJ, USA: Prentice Hall (2003).
15. C. Kelley, Iterative methods for optimization, Philadelphia, USA: SIAM (1999).
16. J. Coplien, D. Hoffman and D. Weiss, Commonality and variability in software engineering, IEEE Software, vol. 15, no. 6, pp. 37-45 (1998).
17. E. Gamma, R. Helm, R. Johnson and J. Vlissides, Design patterns: elements of reusable object-oriented software. Massachusetts, USA: Addison-Wesley, Massachusetts (1995).
18. J. Dennis and R. Schnabel, Numerical methods for unconstrained optimization and nonlinear equations, Philadelphia, USA: SIAM (1996).
19. J. Nocedal and S. Wright, Numerical optimization, 2nd. Springer-Verlag (2006).
20. R. S. Barrera, Arquitectura de Software Flexible y Genérica para Métodos del tipo Newton, Master's thesis, Instituto Nacional de Astrofísica, Óptica y Electrónica, Mexico (2011).

Method for Lightening Software Processes through Optimizing the Selection of Software Engineering Best Practices

Mirna Muñoz, Jezreel Mejia, Juan Miramontes

Centro de Investigación en Matemáticas, Av. Universidad no 222, 98068 Zacatecas, México

{mirna.munoz,jmejia,juan.miramontes}@cimat.mx

Abstract. One of the most important concerns in software industry is the development of software products with the optimal use of resources. Therefore, a software development organization needs to be efficient and have software engineering practices that allow it to be efficient according to its particular needs and business goals. A feasible way to achieve this is by implementing software process improvements that allow the organization the selection and the implementation of those best practices that help it to improve its processes and to become more efficient. This research proposes a method to lighten software processes that enable organizations, especially SMEs, to assess the activities of their software development processes, so that the processes are optimized based on practices that add more value to them, and therefore lead them to be more efficient.

Keywords: software process lightening, software process improvement, added value analysis, agile development practices.

1   Introduction

The software process improvement (SPI) is a planned, managed and controlled effort, which aims to increase the capacity of organizations' development processes [1]. To help organizations to achieve a process improvement, a set of process improvement models and standards have been proposed such as: the Capability Maturity Model Integration (CMMI) [2], and the ISO/IEC 15504 [3] standard, which provide an approach based on a set of best practices that have proved a high performance and success in software organizations [4].

However, these models and standards, most of the times, are not understood in an adequate way, so they are perceived for many organizations as a "heavyweight", that means that are too hard to understand and to implement [5, 6], because of the great amount of resources and the short and long-term commitment that these models and standards require.

Therefore, the required time and costs make difficult to start and to carry out activities regarding the assessment and improvement processes in small and medium enterprises (SMEs)[5, 6, 7].

© Springer International Publishing AG 2017

J. Mejia et al. (eds.), *Trends and Applications in Software Engineering*, Advances in Intelligent Systems and Computing 537, DOI 10.1007/978-3-319-48523-2_11

This research aims to present a method that allows organizations to lighten their software processes through optimizing the selection of software engineering practices and the software tool developed to implement the method.

The rest of the paper is organized as follows: section 2 shows the research background; section 3 presents the proposed method; section 4 shows the case study performed to validate the method; and section 5 presents the conclusions.

2   Background

Software process improvement allows any type of organization, to solve key issues regarding the reduction or elimination of quality problems. Besides, it allows them to examine in an objective way its processes to determine in which level the processes are tying with their specific needs.

Based on the above mentioned, this research work arises from the premise that quality models and standards are not understood in an adequate way by organizations, therefore, they are perceived for many organizations as a "heavyweight".

This highlights the problems that face the organizations in performing activities regarding the processes assessment and improvement within adequate time and costs. Then, the goal of this research work is supporting software organizations in a continuous improvement. So, their processes are optimized through lightening them and without losing the use of best practices with added value for them.

To achieve that, this research work began performing a systematic literature review to establish the state of art for lightening software processes, focusing on three aspects: (1) frameworks, methods and methodologies; (2) targeted processes; and (3) strategies. The complete report was published in [8]. Next, the main results of the systematic review are listed:

1. *Frameworks, methods and methodologies*: Not only frameworks, methods and methodologies for lightening processes were found; in addition, other proposals such as tools and processes were identified. However, most of the proposals are focused on providing frameworks and tools.
2. *Targeted processes*: The model mentioned in most of the studies for lightening processes is the Capability Maturity Model Integration for Development (CMMI-DEV). Besides, the most targeted processes are included in the maturity level 2 and 3, of the engineering and project management categories.
3. *Strategies*: the most used strategy is based on using software tools for automating the process. This strategy allows organizations reducing the time and effort for the process execution, but does not analyze the processes, so it is not possible to get more efficient processes according to the organization needs. Finally, related to the analysis of other strategies, it was concluded that strategies could be integrated or combined for analyzing and at the same time lightening the processes in a proper way.

Taking into account the previous results, it was developed a method for lightening software processes that enables organizations, especially SMEs, to rate the activities of their software development processes, such that, the processes are optimized based on practices that add more value to the processes.

3   Method for lightening software processes

The proposed method is based on the premise that lightening software processes is a way that software development organizations can follow to get efficient processes and to promote a culture of continuous improvement.

The method is based on three strategies: 1) identification of the organizational best practices; 2) use of the lean principle: "added value"; and 3) a combination of best practices provided by formal models, which are named in this research as "formal practices" and agile practices. By this way, it is possible to lighten software processes by optimizing the selection of practices to be implemented to achieve more efficient processes.

It is important to mention that the method was designed for both software organizations with experience in implementing software process improvement and organizations without experience. The method and the tool defined are briefly described in the next subsections.

3.1   Method description

The method consists of five phases, as shown in the Figure 1.

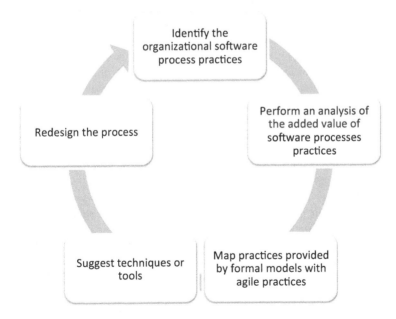

Fig. 1. Method phases

In order to achieve a continuous improvement the phases should be executed in a continuous way and in an iterative manner according to the organization's needs.

The method considers two types of roles: (1) the *process manager,* which can be a member of the software process engineering group or a process manager. He/she is in charge of modeling the organization's process; (2) the *process users,* which are required to identify practices and evaluate them. Both roles have a key participation in the first two phases of the method. Table 1 shows the roles and their responsibilities.

Table 1. Method roles and responsibilities by phase

Phase	Role	Activities
Identify the organizational software process practices	Process manager	Answer questionaries
		Model the process
	Process user	Answer questionaries
Perform an analysis of the added value of software	Process manager	Rate practices
practices	Process user	Rate practices

1. *Identify the organizational software process practices.* In this phase, the practices of the process are characterized. So, the process is modeled with base on the current practices implemented in the organization. To identify the practices, this phase uses a questionnaire, which provides a set of formal practices and aims to know which practices are implemented in the organization. Then, a diagram of the process, according to the identified formal practices, is made.
2. *Perform an analysis of the added value given to the software practices.* This phase analyzes the added value for each formal practice. First, should be collected a set of rates for each practice assigned by a set of process users. Then, an analysis of the rates of each practice is performed in order to classify the formal practices according to four types: (1) those practices that add a high value to the process; (2) those practices that add medium value to the process; (3) those practices that add minimum value to the process; and (4) those practices that do not add value to the process. Figure 2 shows the corresponding action that can be performed according to the classification of each practice.
3. *Map formal practices with agile practices.* This phase uses a mapping between practices of CMMI and Scrum based on the mapping works of Marçal [9] and Díaz [10]. Besides, the mapping performed in this method provides a set of guidelines for the implementation of each agile practice.
4. *Suggest techniques and tools.* In this phase, a set of tools and techniques is suggested for each agile practice to facilitate its implementation. So that, it can be possible to facilitate its implementation, and therefore to increase its value.
5. *Redesign the process.* The last phase consists on redesigning the process based on three recommendations: (1) removing the practices that do not add value and do not have a high impact toward the achievement of a goal; (2) changing a formal practice by an agile practice and (3) selecting a technique or a tool from the set of suggested tools and techniques in order to facilitate the practice implementation and/or execution.

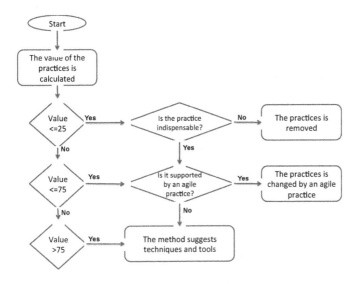

Fig. 2. Flowchart of the actions to be performed according to the value of the practice

3.2   Tool for using the proposed method

This section provides an overview of the web tool that supports the use of the method for lightening software processes.

The tool starts providing a screen in which a user can register an account or sign in. Besides, data such as the number of the employees and the user role are required.

Then, the main screen of the tool is presented. As Figure 3 shows, the main screen contains a set of question with examples of a set of products obtained after performing the practice this allows identifying the organization's practices as well as rating those practices, which means, the added value to the practice. Finally, at the bottom of the page a status bar is shown.

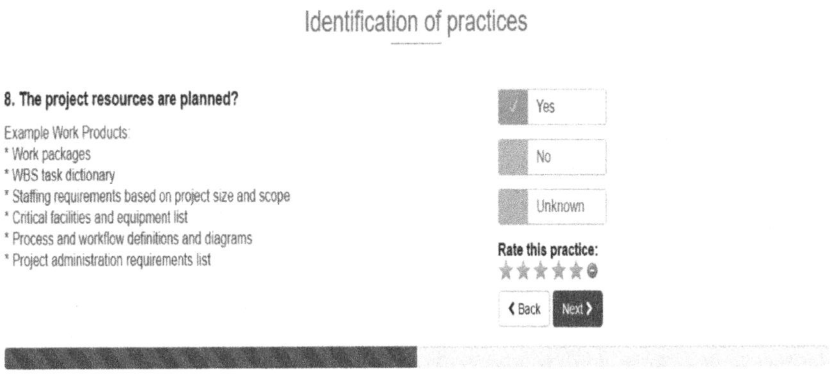

Identification of practices

8. The project resources are planned?

Example Work Products:
* Work packages
* WBS task dictionary
* Staffing requirements based on project size and scope
* Critical facilities and equipment list
* Process and workflow definitions and diagrams
* Project administration requirements list

Yes

No

Unknown

Rate this practice:
★ ★ ★ ★ ★

‹ Back   Next ›

Fig. 3. Screen to identify and rate practices

When the user is a process manager, she or he models the organization's process. The modeling is done using a web-based tool BPMN diagrams powered by bpmn.io [11]. Figure 4 shows the interface to model the organization's process. The process practices are automatically added while the user answers affirmatively to questions. Besides, the process manager can customize the diagram to reflect the current process.

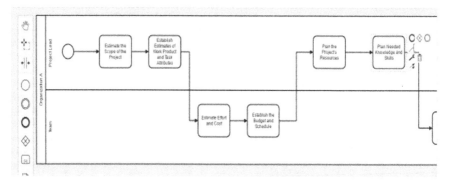

Fig. 4. Interface to model the software process

4   Case Study

To validate the viability of the method a case study was performed. Next, the case study design and the results are presented.

4.1   Case study design

Following the case study guidelines provided by Wieringa in [12], the case study design was established as follows:

1. *Knowledge goal*: the goal of this case study was to validate the viability of the method.
   Current Knowledge: Lighten software process is a way that software organization can follow to make efficient its processes promoting an improvement processes culture. Having less heavy processes allows the organization to use formal best practices, which satisfy customer needs with less cost and promotes quality. Besides, according to the Chaos Manifesto [13], two factors that allow organization to achieve successful software development projects are the optimization and the agile processes.
2. *Knowledge questions*: Table 2 shows a set of the knowledge questions and its related assessed goal.
3. *Population*: The case study was focused on process managers and process users of a software development SME.
4. *Measurement Design:* (1) *variables*: identified practices, agile practices suggested, and tool and techniques suggested; (2) *data source*: process managers, and process users; (3) *measurement instruments*: via web survey using Google documents; (4)

measurement schedule: the process manager and process user according to a sample suggested by the tool use the tool. After a survey had to be answered.

Table 2. Knowledge questions and their assessed goal

#KQ	Knowledge Questions	Assessed Goal
1	Does the method allow identifying the organization's process practices?	This question aims to validate if the method allows identifying the current organizational practices, best practices and those that do not add value to the process.
2	Are the agile practices suggested by the method adequate?	This question aims to validate if the suggested agile practices are perceived as adequate to be implemented in the organization
3	Are the tools and techniques suggested by the method adequate?	This question aims to validate if the suggested tools and techniques are perceived as adequate to be implemented in the organization.

4.2 Case study execution

The SME develops informatics' solutions and systems and has 50 employees. Besides, it offers implementation, training and consulting. It has experience in process use and improvement. To execute the case study, the tool was used by 12 employees according to the sample suggested by the tool given the number of organization's employees.

The case study started by introducing the participants to the case study goal. Then, each participant used the web tool. Finally, each participant answered a survey via web, defined to assess the method viability as well as the developed tool.

4.3 Case study results

This section shows the results of the case study. The results obtained by the SME are shown in Figures 5, 6, 7 and 8 as follows:

Figure 5 shows the results of the practices classifications. As figure shows, all practices are classified between the range of high and medium added value. This is because the organization has experience in the use and improvement of processes.

Analysis of the value added ▲

ID	Practice	Value	Range
1	Estimate the Scope of the Project	51.67	51-75
2	Establish Estimates of Work Product and Task Attributes	58.33	51-75
3	Define Project Lifecycle Phases	83.64	76-100
4	Estimate Effort and Cost	58.33	51-75
5	Establish the Budget and Schedule	58.33	51-75
6	Identify Project Risks	51.67	51-75
7	Plan Data Management	46.67	26-50
8	Plan the Project's Resources	41.67	26-50
9	Plan Needed Knowledge and Skills	86.67	76-100
10	Plan Stakeholder Involvement	32.73	26-50
11	Establish the Project Plan	84.00	76-100
12	Review Plans That Affect the Project	50.00	26-50
13	Reconcile Work and Resource Levels	46.67	26-50
14	Obtain Plan Commitment	58.33	51-75

Fig. 5. Practices classifications

Figure 6 shows the results of a set of agile practices according to the organization's practices identified. As figure shows the screen includes an indicator of the coverage that the agile practice provides to the organization's practices. Besides, a guide of how to implement the agile practice is provided.

Mapping to agile practices ▲				
ID	Practice	Support	Scrum Practice	Guidelines
1	Estimate the Scope of the Project	▢	Unsupported	Show guides
2	Establish Estimates of Work Product and Task Attributes	■	Estimate attributes using Planning Poker or Magic Estimation	Show guides >
4	Estimate Effort and Cost	▣	Estimate effor of the user stories	Show guides >
5	Establish the Budget and Schedule	▣	Establish Sprints	Show guides >
6	Identify Project Risks	▣	Identify impediments list	Show guides >
7	Plan Data Management	▢	Unsupported	Show guides
8	Plan the Project's Resources	■	Establish the Scrum Team	Show guides >
10	Plan Stakeholder Involvement	▣	Define roles and responsibilities at the beginning and end of the Sprint	Show guides >
12	Review Plans That Affect the Project	■	Review plans that affect the project in the Sprint Review	Show guides >
13	Reconcile Work and Resource Levels	■	Reconcile work and resource levels in the Sprint planning	Show guides >
14	Obtain Plan Commitment	■	Obtain plan commitment in the Sprint planning	Show guides >

Fig. 6. Agile practices suggested and level of coverage

Figure 7 shows the results of a set of techniques and tools according to the organization's practices identified. The set of techniques and tools are suggested to support the organization in the implementation or use of the practice. It is important to mention that the tools and techniques are provided to those practices that have high, medium and low added value in order to help organization to improve the added value of the practice.

Suggested techniques and tools ▲	
ID Practice	Technique or tool
1 Estimate the Scope of the Project	• WBS templates of previous projects • Historic information • Decomposition: subdivision of deliverables and subentregables
3 Define Project Lifecycle Phases	• Subdivide the project into smaller, more manageable components to support project activities (planning, execution, control and completion) • WBS template can be used as templates for new projects
7 Plan Data Management	• Communication skills • Manual filing systems • Electronic databases • Software of project management • Standardized forms or templates
9 Plan Needed Knowledge and Skills	• Team building activities • General management skills • Rewards and recognition systems • Disposition • Historic information • Staff and new hires • Internal training • External training • Acquisition of external skills
11 Establish the Project Plan	• Project planning methodology • Knowledge and skills stackeholders • Project Management Information System (PMIS) • Earned Value Management (EVM) • Standardized forms or templates • Simulations

Fig. 7. Set of suggested tools and techniques

Figure 8 shows the results of the redesigned process according to the recommendations of the method. As figure shows the new practices graphic provides tags, which indicate that it is suggested to change the organization's practice with an agile one.

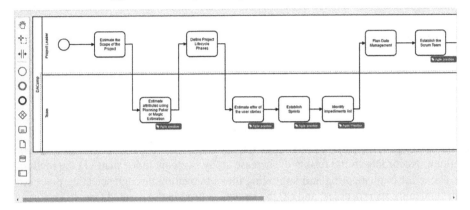

Fig. 8. Redesigned process

Finally, Figure 9 shows the results of the variables defined in the case study focused on validating the method.

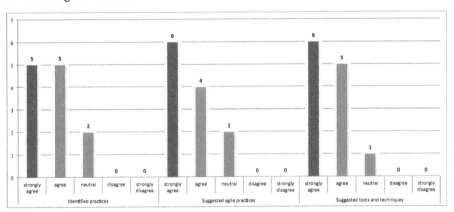

Fig. 9. Data collected from three main variables through the web survey

The analysis of the main variables:

- *Identified practices*. This variable allows getting information to answer the *KQ1* (see Table 2). *Does the method allow identifying the organization's process practices?* As Figure 9 shows, 5 engineers answered "strongly agree" and 5 "agree", which means that they considered the method allowed identifying the current organizational practices, best practices and those that do not add value to the process.
- *Suggested agile practices*. This variable allows getting information to answer the *KQ2* (see Table 2). *Are the agile practices suggested by the method adequate?*

As Figure 9 shows, 6 engineers answered "strongly agree" and 5 "agree", which means that they considered the agile practices suggested by the method adequate to be implemented in the organization.

- *Suggested tools and techniques.* This variable allows getting information to answer the *KQ3* (see Table 2). *Are the tools and techniques suggested by the method adequate?* As Figure 9 shows, 6 engineers answered "strongly agree" and 5 "agree", which means that they considered the tools and techniques suggested by the method adequate to be implemented in the organization.

5   Conclusions

This paper presents a method that aims to help organizations in lightening their process as a way to perform a continuous improvement and to develop an improvement culture. According to the results of the case study, we concluded that: (1) the method is viable for implementing and improving the organization through selecting practices for lightening its processes; and (2) the strategies used in the method (identification of best practices, use of the lean principle "added value" and the combination of formal practices and agile practices) are adequate for lightening software processes.

Besides, the validation of the method demonstrates that the method and its tool have a good acceptation for the participants because they perceived that the use of the method could help them to increment the efficiency of their process. As future work, we want to improve the sample of organizations to get more results about the method in different contexts and type of organizations.

References

1. Pino, F. J., Baldassarre, M. T., Piattini, M. and Visaggio, G. Harmonizing maturity levels from CMMI-DEV and ISO/IEC 15504. Journal of Software Evolution: Research and Process., 22(4): 279–296. (2010),
2. CMMI Product Team: CMMI® for Development, Version 1.3. Pittsburgh, PA (2010).
3. ISO/IEC: ISO/IEC 15504 Information Technology—Process Assessment (Parts 1–5) (2003)
4. Pettersson, F., Ivarsson, M., Gorschek, T., Öhman, P.: A practitioner's guide to light weight software process assessment and improvement planning. Journal of System and Software. 81(6), 972–995 (2008).
5. Kuilboer, J., Ashrafi, N.: Software process and product improvement: an empirical assessment. Information and Software Technology 42(1), 27–34 (2000).
6. Reifer, D.J.: The CMMI: it's formidable. Journal of System and Software. 50(2), 97–98 (2000).
7. Calvo-Manzano, J.A., Cuevas A., San Feliu, T., De Amescua, A., Sánchez, L., Perez-Cota, M: Experiences in the application of software process improvement in SMES. Software Quality Journal. 10, 261–273 (2002)
8. Miramontes J., Muñoz M., Calvo-Manzano J.A. and Corona B.. Establishing the State of the Art of Frameworks, Methods and Methodologies Focused on Lightening Software Process: A Systematic Literature Review. Springer International Publishing Switzerland 2016

J. Mejia et al. (eds.), Trends and Applications in Software Engineering, Advances in Intelligent Systems and Computing 405. (2015).

9. Marçal A., De Freitas B., Furtado F.S., and Belchior A.D., Mapping CMMI project management process areas to SCRUM practices, in *Proceedings of the Internatinal Conferece on Software Engineering*, pp. 13–22, (2007).

10. Diaz J., Garbajosa J., and Calvo-Manzano J.A., Mapping CMMI Level 2 to Scrum Practices: An Experience Report, *Software Process Improvement*, vol. 42, pp. 93–104, (2009).

11. Camunda Services GmbH, BPMN 2.0 rendering toolkit and web modeler. bpmn.io, (2016). [Online]. Available: https://bpmn.io/.

12. Wieringa R.J. "Observational Case Studies" in Design Science Methodology for information Systems and Software Engineering, Springer-Verlag Berlin Heidelberg 2014. pp. 225-245. (2014).

13. The Standish Group International. CHAOS MANIFESTO 2013: Think Big, Act Small. *The Standish Group International*, 1–52. Retrieved from http://www.standishgroup.com. (2013).

Scheme for the automatic generation of directions to locate objects in virtual environments

Graciela Lara[1], Angélica De Antonio[2], Adriana Peña[1], Mirna Muñoz[3], Luis Casillas[1]

[1] CUCEI of the Universidad de Guadalajara,
Av. Revolución 1500, Col. Olímpica, 44430 Guadalajara (Jalisco), México.
graciela.lara@red.cucei.udg.mx, {adriana.pena, luis.casillas}@cucei.udg.mx
[2] Escuela Técnica Superior de Ingenieros Informático of the Universidad Politécnica de Madrid,
Campus de Montegancedo, 28660 (Boadilla del Monte), España.
angelica@fi.upm.es
[3] Centro de Investigación en Matemáticas
Avda. Universidad no. 222, 98068, Zacatecas, México
mirna.munoz@cimat.mx

Abstract. The automatic generation of direction in natural language, for the location of objects, is an ongoing research area heavily supported by the use of virtual environments (VEs). Important components of spatial language such as the selected reference object, along with specific features related to the situation of the scenario and the user, have to be properly combined in order to create a helpful direction to locate an object within a VE. In this paper we present a scheme, constructed upon literature review and specific empirical data, to link those different elements related to the location of objects, aimed to establish the suitable algorithms for the automatic generation of spatial language in VEs.

Keywords: virtual environments, location of objects, spatial language.

1. Introduction

Humans, animals and objects occupy a place in space; in consequence, we have developed spatial knowledge, a basic skill for the localization process. Albeit a seemingly simple task, spatial language to express where objects are located, calls for a mix of spatial knowledge and an accessible visual representation for our linguistic system [1].

The automatic generation spatial language, directions for the location of objects, represents a challenge with a number of difficulties. Take for example, the use of absolute references that might cause confusion with respect to relative references [2]; or the fact that there is not a straight

J. Mejia et al. (eds.), *Trends and Applications in Software Engineering*, Advances in Intelligent Systems and Computing 537, DOI 10.1007/978-3-319-48523-2_12

forward method to select a reference object [3]. Even though, a number of applications applying spatial language have been developed, among which are included: graphic design and drawing programs, computer games, navigations aids, robot systems, training simulators and geographic information system interfaces [3], where virtual environments (VEs) play an important role.

1.1.   Related Work

Concerning VEs, in his doctoral thesis, Kelleher [4] developed a computational framework, perceptually based, for the interpretation of spatial language. The computer system is a VE for a user to navigate and manipulate objects. It contains a mechanism for the user to select different frames of reference thorough a semantic framework with locative prepositions. On it, a visual saliency algorithm is used for the integration of the speech.

Also for the interpretation of spatial language, Gorniak & Roy [5] developed a system through a model that describes objects of 3D scenes but with spatial relations interpreted in 2D. Their system has an algorithm to extract visual features of the objects, and it manages the description of spatial relations. Their descriptive spatial language contains hundreds of reference-expressions based on similar scenes, a syntactic analyzer of spoken expressions, and a composition engine managed by an interpreter with various lexical units.

For the automatic generation of spatial language in VEs, Barclay [3] developed a model for processing scene descriptions that operates in realistic environments, tested on a large set of 3D scenes representations. This project emphasizes the use of references and spatial relations for locating objects, through perceptual salience.

More recently, Trinh [6] developed a system with a tool for semantic modeling of spatial relations among objects in VEs, where the spatial relations are specified at a conceptual level. The model focuses on the spatial limitations of VEs, such as space communication difficulties. For a detailed analysis of computer systems with spatial language see Lara, De Antonio & Peña [7].

None of these systems incorporates the user modeling. In this paper we present a scheme that deals with the complexity and specific elements that are combined for the automatic generation of spatial language. The scheme includes an original algorithm for 3D objects, to select the best reference object; and it also, incorporates significant aspects of the user modeling, to give direction to a user in a VE for the location of objects.

2.   Elements for the automatic generation of spatial language

As mentioned, spatial language involves different concepts and elements. Because the location of an object is inherently relative, to place it, a *frame of reference* has to be established. The spatial frames of reference provide a structure to specify the object's spatial composition and position; a coordinate system to give directions from different points in space or a mental representation of positions such as up, down or side [8]. In the adoption of a specific frame of reference, an object can be pointed in relation to: the observer, the environment, its own intrinsic structure, or other objects in the environment [9].

A common practice to give direction for the location of an object is the use of *reference objects*; in fact, it might be difficult to state the position of an object without referring to another [8]. The selection of a *reference object (RO)* conveys the recognition of its prominent features, that is, its *perceptual saliency*. Early mentioned by Titchener [10], this key concept for the location of objects has been described as those features of the object that somehow draw our attention [11, 12, 13], mainly probable because they are rare or just different in the scenario [14,15]. However, as Gapp [16] stated out, in some cases, the selection might obey only to the distance between the *object to be located (OL)* and the RO. The use of a RO implies to establish a spatial relation between the OL and the RO, see Figure 1.

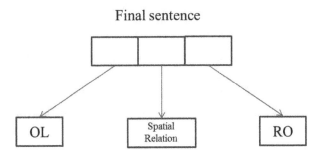

Fig. 1. Structure of a spatial language sentence

In addition, the good perception of an object depends on factors such as visual acuity, clarity of vision and the viewpoint of the observer, as well as on the social and psychological impact that these features might have on the viewer.

A first approach to give directions to a person for the location of an object is to take into account if the OL is within the *field of view* of that person [17]. If the OL is not in the users' field of view, it might or not be the case that they are in the same room (or space). In both cases, extra directions are required to place the user in a position in which his/her field of view reaches the OL; and both are out of the scope of this paper.

The scheme here presented is then focused to the particular case that the OL is within the user's field of view. This leads to diverse situations, from which were included in the scheme the next cases: A) The object is salient by itself; B) The object is somehow occluded; and C) The object is not salient and therefore a RO is required.

The elements for the automatic generation of spatial language situations will be next described, and in the next sections how to use those elements for the three aforementioned situations, including others derived from them, are discussed.

2.1.   Syntactic structure of the spatial language

Kelleher [4] proposed a linguistic structure proper for the automatic generation of spatial language, the one shown in Figure 2. In which the syntactic structure is divided into a nominal syntagm that refers to the OL, and a verbal syntagm and adverbial syntagm. The nominal syntagm includes the article "The", + an optional (Object_feature) + the OL; an example is: "The yellow pen". The verbal sytagm specifies the nature of the problem, in this case to indicate the positional situation, which is always "is". Finally the adverbial syntagm contains both the spatial relation and the RO; its structure comprehends the "spatial relation" + the article "the" + an optional (Object_feature) of the RO + the RO. It is worth to mention that the adverbial sytagm changes when it includes several references object, take for example this adverbial syntagm: "...in front of the desk, between the red ball and the bicycle, to the right of the second printer."

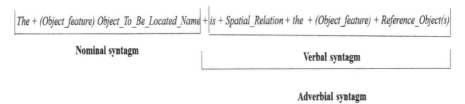

$$\underbrace{The + (Object\_feature)\ Object\_To\_Be\_Located\_Name}_{\textbf{Nominal syntagm}} + \underbrace{is + Spatial\_Relation + the\ + (Object\_feature) + Reference\_Object(s)}_{\textbf{Verbal syntagm}}$$

Adverbial syntagm

Fig. 2. Syntactic structure of a spatial language sentence

In order to generate in an automatic form this structure, we need to establish a reference object and then select the proper spatial relation.

2.2.   Algorithm for the Selection of the Best Reference Object

Three criteria were considered for this algorithm: *visual saliency* of the object, and from the user: *prior knowledge* [16, 18], and the *probability to remember*. The visual saliency of an object might relay on a number of factor. However, because VEs are predominantly visual and based on a literature review, we proposed in [19], the use of the more prominent objects' features related to the human vision, that is: color, size and shape.

For the automatic interpretation of these features in a VE, in [20] we proposed a weighted combination of these three measures of the features of the objects, and how to get them, in order to calculate the saliency by each object in the scenario, to obtain an ordered relation by perceptual saliency of all of them.

The second criterion for the algorithm is the prior knowledge of the user. It refers to the areas of knowledge to which the objects belong and the user's probability to posses them, based on his/her previous training or experience. It represents the level of familiarity that a user possesses with a certain area of knowledge and the objects associated with this area. For example, a computer engineer should have certain knowledge regarding computer input/output devices. This requires for the system to manage a user modeling [21], with a test from which a degree of familiarity related to the analyzed object can be established.

The last criterion for the algorithm is the probability to remember, that is, the probability for a user to remember a previously seen object in the VE. For it, three factors can be considered: the perceptual saliency of the object in question, the user's ability to remember objects' location and the history of the user visibility of the object within the VE.

The first factor, the perceptual saliency of the object, can be calculated as in [20]. For the user's probability to remember the location of an object, the Rey-Osterrieth complex figure text [22] provides a scale for each particular user, which in turn will be part of the user modeling.

As of the history of vision, it involves several concepts: the maximum viewed surface of the object inside the scene from any of the past user's points of view, the clarity of vision or clarity of perception, the time of vision exposure of an object by the user, and the time of oblivion (the involuntary action of stopping remembering or keeping in memory the information of the objects observed in a scene). All of which imply a number of calculation, regarding the history of the different user's fields of view during his/her navigation in the VE.

This algorithm will proportionate the best RO, and by including in it the OL, it will also be able to establish if the OL is salient by itself.

A third part of the syntactic structure (see Figure 1) is the spatial relation. An algorithm to select it is next described.

2.3.   Algorithm for the Selection of the Spatial Relation

Gapp [23, 24] divided spatial relations into two classes: topological, those that refer to a region proximal to an object (e.g. "at", "near"); and projective, that take into account the relation between those objects (e.g. "in front of", "behind" or "above"). The relation "between" has an exceptional position in the group, because it refers to two objects, its basic meaning is defined by the structure of its region of applicability; the location with the highest degree of its applicability is the midway between two ROs.

Different criteria for the automatic choice of a particular spatial relation can be applied. A set of selected relations and the criteria to use them is here briefly described. The set includes some of the most commonly used, and that we consider cope most of the situations.

On/Under are spatial relations suitable when the OL is in contact with the RO, therefore a collision in the VE has to be detected, in the "Y" axis.

Close to is a proper relation when the distance between the OL and the RO does not exceed certain threshold.

Left/Right/Above/Below/In front of/Behind are spatial relations mutually exclusive. The key to choose one is to establish, from the point of view of the user, the nearest objects to obtain their roominess. The roominess will evaluate the objects' points that fall down inside or out of the distance, from the user to the objects in his/her field of view.

Inside requires first determining if the OL is inside another one. In this case can be applied the ray-casting technique; by evaluating the hit object or objects through several rays directed from the center inside of an object, it can be inferred an outsider object.

Between can be used once is evaluated if there are several ROs candidates. The two closest objects of reference (RO1 and RO2) are then identified and their distance to the OL is calculated. A criterion based on the object with highest distant, lower than the nearest distance to the OL with a threshold difference, can be applied to use this spatial relation.

Once the best reference objects and a spatial relation are established then the next step is to generate spatial language, this algorithm is next described.

3. Algorithms to Generate Spatial Language

Let us now have a close look to the syntaxes for the different mentioned situations, when the OL is in the field of view of the user.

A) **The OL is salient on itself**; this situation does not require a RO to generate its location direction. In this case, for simplicity, we proposed to use the user as frame of reference, which means that the word "you" will be used instead of the RO. The structure is then (as described in section 2.1):

"The", + an optional (Object_feature) of the OL + the OL + "is" + the "spatial relation" + "you".

Example: "The yellow box is in front of you"

B) **The OL is somehow occluded**; this could be because the object is inside another one, or because another object occludes it. In any case the object that occludes the OL becomes a secondary object to be located (OL2) and a second spatial relation is required to indicate the relation between the OL and the OL2, that in turn requires a secondary RO (RO2). The structure, in this case is:

"The", + an optional (Object_feature) of the + the OL + "is" + the "spatial relation" + "the" + the OL2 + "that is" + "spatial relation" + "the" + an optional (Object_feature) of the RO2 + the RO2

Example: "The blue ball is inside the white box that is on the brown desk"

C) **The OL is not salient and it requires a RO**, in this case the RO might or not have a high probability of being remembered by the user, both cases are treated separately. In the case that there is a high probability that the user remembers a RO, inside a certain radius near to the OL, then the spatial relation between them is determined to generate the directions with the same structure used in the A) situation.

If there is not a high probably for the user to remember any of the ROs within a radius of the OL, then the perceptual saliency of the objects is applied to them, and the sentence with directions is structured as in the situation A). These situations are summarized in the next Table 1.

Table 1. Criteria for different situations for the automatic generation of directions.

The OL is in the field of view of the user	
Situations	Criteria
A) The OL is salient on itself	• The directions include a spatial relation of the intrinsic type (listener centred).
B) The OL is somehow occluded	• The directions consider the object that contains the object to be located, or the object that occludes it, and it is transformed into a secondary object to be located OL2.
C) The OL is not salient and it requires a RO	• In this situation is considered the users' probability to remember the RO.
C.1) The RO has a high probability of being remembered by the user	• The direction is generated, considering the RO with the maximum probability of being remembered. • The user has previous knowledge about the RO. • The spatial relation is determined between the OL and the RO.
C.2) The RO does not have a high probability of being remembered by the user	• The RO selected is the one with maximum value of saliency, the highest probability of being remembered by the user, and the user's prior Knowledge.

4.   Conclusions and Future Work

The automatic generation of spatial language is a complex task that requires the incorporation of a number of factors. All the objects' features in a VE can be explored by the computer system to categorize them by their saliency and their proximity to the user. It can also be established the field of view of the user at each moment of his/her navigation in the scenario. And, it can be included a user modeling to personalize, to some extend, the spatial language. A number of elements that require schematization, in order to automatically generate a proper direction for the user to locate an object in a VE. In this paper we proposed such a scheme, including the algorithms to select a better reference object, to select a proper spatial relation and to generate the spatial language.

Other features can be included in the future to this scheme. Regarding the user modeling, cognitive and emotional perceptions that might probably have an influence in the selection of a reference object. It can be also included other visual features of the objects, for example its texture. Finally, the directions to place the object in the field of view of the user can be included.

References

1. Landau B, Jackendoff R: "What" and "Where" in Spatial Language and Spatial Cognition. *Behavioral and Brain Sciences* 1993, 16(217 - 265).
2. Pederson E, Danziger E, Wilkins D, Levinson S, Kita S, Senft G: Semantic Typology ans Spatial Conceptualization *Linguistic Society of America JSTOR Language* 1998, 74(3):557 - 589.
3. Barclay M: Reference Object Choice in Spatial Language: Machine and Human Models. University of Exeter. PhD. Thesis; 2010.
4. Kelleher JD: A Perceptually Based Computational Framework for the Interpretation of Spatial Language Dublin: Dublin City University. PhD. Thesis; 2003.
5. Gorniak P, Roy D: Grounded Semantic Composition for Visual Scenes. *Journal of Artificial Intelligence Research* 2004, 21:429 - 470.
6. Trinh T-H: A Constraint-based Approach to Modelling Spatial Semantics of Virtual Environments. Université de Bretagne Occidentale. PhD. Thesis; 2013.
7. Lara G, De Antonio A, Peña A: Computerized spatial language generation for object location. *Springer Virtual Reality* 2016:1 - 10.
8. Mou W, McNamara TP: Intrinsic Frames of Reference in Spatial Memory. The American Psychological Association Journal of Experimental Psychology: Learning, Memory, and Cognition 2002, 28(1):162 - 170.
9. Wraga M, Creem SH, Proffitt DR: The influence of spatial reference frames on imagined object and viewer rotations *Elsevier Acta Psychologica* 1998, 102:247 - 264.
10. Titchener EB: Lectures on the Elementary Psychology of Feeling and Attention. In. New York: The MacMillan Company; 1908.

11. Frintrop S, Rome E: Computational Visual Attention Systems and their Cognitive Foundations: A Survey. *ACM Journal Name* 2010, 7(1):1 - 46.
12. Vargas ML, Lahera G: "Asignación de relavancia": Una propuesta para el término inglés "salience". In: *Actas Esp Psiquiatría.* vol. 39. España; 2011: 271 - 272.
13. Lahera G, Freund N, Sáin-Ruíz J: Asignación de relevancia (salience) y desregulación del sistema dopaminérgico. *Elsevier Doyma Revista de Psiquiatría y Salud Mental* 2013, 6(1):45 - 51.
14. Hall D, Leibe B, Schile B: Saliency of Interest Points under Scale Changes. In: *British Machine Vision Conference (BMVC'02).* Cardiff, UK; 2002: 646 - 655.
15. Röser F, Krumnack A, Hamburger K: The influence of perceptual and structural salience. In: Cooperative Minds: Social Interaction and Group Dynamics In Proceedings of the 35th Annual Meeting of the Cognitive Science Society: 2013; Austin, TX. USA; 2013: 3315 - 3320.
16. Gapp K-P: Object Localization: Selection of Optimal Reference Objects. In. *Fed. Rep. of Germany: Universität des Saarlandes* 1995: 1-18.
17. Harrington DO, Drake MV: Los campos visuales: texto y atlas de perimetría clínica. In.: Ediciones Científicas y Técnicas; 1993.
18. Gapp K-P: Selection of Best Reference Objects in Objects Localizations. In: In Proceedings of the AAAI Spring Symposium on Cognitive and Comutational Models of Spatial Representations: 1996b; Stanford, CA.; 1996b: 23 - 34.
19. Lara G, Peña A, De Antonio A, Ramírez J, Imbert R: (in press) Comparative analysis of shape descriptors for 3D objects. *Multimedia Tools Applications* 2016a:1 - 48.
20. Lara G, De Antonio, Peña A: A computational model of perceptual saliency for 3D objects in virtual environments; 2016b (in review).
21. Kobsa A: Generic User Modeling Systems. *Springer User Modeling and User-Adapted Intraction* 2001, II:49 - 63.
22. Osterrieth PA: Le test de copie d'une figure complexe. *Arch Psychol* 1944, 30:206-356.
23. Gapp K-P: From vision to language: A cognitive approach to the computation of spatial relations in 3D space. In., Künstliche Intelligenz - Wissensbasierte Systeme. Bericht Nr. 110 edn. Fed. Rep. of Germany: Universität des Saarlandes; 1994: 1 - 19.
24. Gapp K-P: Angle, distance, shape, and their relationship to projective relations. In: In Proceedings of the 17th Annual Conference of the Cognitive Science Society: 1995a; Cognitive Science Society Mahwah, NJ.; 1995a: 112 - 117.

Factors Affecting the Accuracy of Use Case Points

Luis Morales Huanca, Sussy Bayona Oré

Unidad de Posgrado de la Facultad de Ingeniería de Sistemas e Informática, Universidad
Nacional Mayor de San Marcos (UNMSM). Av. Germán Amézaga s/n, Lima, Perú
luis.morales2@unmsm.edu.pe, sbayonao@hotmail.com

Abstract. The success of a software development project depends on that the
retrieved product complies with the user's specifications and to be completed
within the time and within the budget established. Many projects fail when they
are not being developed within the time set due to a bad assessment of the effort
or duration of the software project. In this article is presented the results of a
literature review about the factors that affect the precision of the use case points
method. A total of 37 primarily studies were selected. The results show that the
environmental factors, the use cases complexity, the lack of use case
standardization, technical factors and counting transactions are some factors
that affect the use of the use case points method.

Keywords: Software development projects; Use Case Points; Effort estimation;
Use cases.

1 Introduction

Effort estimation in software projects is a key aspect in the software industry, because
they frequently suffer from overrun costs [1]. In the software industry, is of the
utmost importance to predict, as early as possible, what will be the required effort in a
software project [2]. Many researchers have used the object-oriented method, also
known as use case points method, in the estimation of the effort required by a
software development project. The result obtained through this method, consists in an
estimation of the development time in a project by allocating a "weight" to a specific
number of factors that affects it. Then, starting from those factors, one can calculate
the total estimated time of the project [3].

The effort estimation problem is for sure one of the main concerns in software
development. In the intent of solving this, different methods have been developed.
However, the accuracy of the estimation is conditioned by several factors. In this
article will do a review of those factors that affect the estimation which uses the use
case points method. For these, it will be taken as reference studies carried out between
the years 2010 to 2016. The Section 2 presents a theoretical background. Then
Section 3 shows a literature review, while Section 4 consists on the report review and,
finally, Section 5 presents the conclusions.

© Springer International Publishing AG 2017
J. Mejía et al. (eds.), *Trends and Applications in Software Engineering*, Advances
in Intelligent Systems and Computing 537, DOI 10.1007/978-3-319-48523-2_13

2 Background

2.1 Effort Estimation

The effort estimation to invest in software development projects is vital to project planning since it gives a measure of the time and cost during the life cycle of a software project. At the same time, it continues to be one of the most difficult task in software project management, although there are techniques that allows to perform it [4]. Estimates are predictions regarding future performance based on available knowledge [5]. Early software size estimation helps projects to be managed efficiently and helps managers to predict the effort, planning and costing the project [6].

In 2011, Mishra, Pattnaik and Mall indicated that the effort estimation would also depend on factors such as software complexity, software size, the level of expertise of the development team and the development tools used [7]. In 2012, Wen, Li, Lin, Hu and Huang proved that the software development effort estimation is the process of predicting the effort required to develop a software system [8]. Sharma and Kushwaha demonstrated that the software development estimation effort is the process of calculating in the most realistic measure the effort required to develop or maintain the software, based on inputs like software requirements, function points, size of proposed software, use case points, etc [9].

On the other hand, Singh and Sahoo revealed that the software estimation effort is the process of predicting the most realistic use of effort required to develop or maintain a software [10]. In 2014 Subriadi, Sholiq and Ningrum illustrated that the software estimation is an activity to predict or forecast the output of a project, to review the planning, cost, risk and the effort in the project [11]. Finally, in 2015 Saroha and Sahu specified that in software development life cycle, effort estimation is responsible for projects being delivered on time. Exists many effort estimation methods, that are suitable on different environment, software, etc., but still they are not always giving optimal results because of some issues or factors [12].

2.2 Use Case Points

Use Case Points is a method used to predict in man-hours the effort that it will take to develop a software project. This method was originally proposed by Gustav Karner, who defines it as a model that gives an estimation of the size of the effort to develop the system, which can be mapped in man-hours, to complete various phases of the objective, or complete the whole project [13].

Use Case Points method has strengths and drawbacks. Some of the strengths are (1) that it can be used in early stages in order to estimate the effort, and this being measured in a model of use cases that defines the functional scope of the developed software system [14]; (2) is versatile and extensible method in different development and testing projects [15]; (3) use case points method has become well known and widely accepted because uses two common practices: the object-oriented paradigm and the use cases to describe functional requirements [16]; (4) the method is characterized by its relative simplicity [17]; (5) it is also easy to apply [10]; and, (6) is very useful to measure the estimation effort, since is using use cases as input [11].

However, Use Case Points method also presents some weaknesses to mention (1) the uncertainty of the cost factors and the abrupt classification of use cases [16], (2) that the level complexity level of use cases and actors classification needs calibration [6], (3) lacks of information about the counting the number of actors and use cases and that the technical and environmental factors are out of date [10], (4) it has a limitation because assumes that the relationship between software effort and size is linear [18] and (5) the first version of Use Case Points method lacks validation and examination about its reliability for software organizations [19]. As a result of the weaknesses of the Use Case Points method, there is several factors that affect the accuracy in the estimation of this method. The study of these factors will be analyzed in more detail in section 4, Report Review.

3 Literature Review

The literature review refers to the structure and rules that make up this review. With this review is to identify the factors affecting the estimation using use case points method. The present review consists of the following steps: identify research questions, design of the search strategy, conduct of the review, data synthesis, results, and discussions. The Figure 1 presents the steps above:

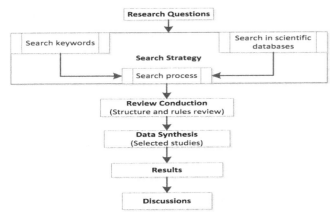

Figure 1: Stages of the literature review

3.1 Research Questions and Search Strategy

The research questions are:

- What factors or limitations affect the precision of the use case points method estimation?
- What are the future work to be carried out to improve the use case points method?

The following sources have been selected to carry out the research process: ACM Digital Library (http://portal.acm.org), IEEE Xplore (http://www.ieee.org/web/publications/xplore/), Science Direct – Elsevier (http://www.elsevier.com), Springer Link (http://www.springerlink.com), Taylor and Francis (www.tandfonline.com), Google Scholar (www.scholar.google.com). Were defined the keywords used to find relevant studies in the titles, summary and contents of articles in the period 2010-2016: "Use Case Point" or "Use Case Points" or "Puntos de casos de usos" or "Puntos de casos de uso" or "Puntos de caso de uso" or "Puntos de caso de usos" or "Punto de casos de usos" or "Punto de casos de uso".

3.2   Conduct of the Review

This section refers to the structure and rules to conduct the review. Listed below are the steps to conduct the review (see Figure 2): (1) was sought in the selected databases to identify relevant studies using the search keywords and search range. (2) Exclude irrelevant studies based on the removal of duplicate studies and analysis of titles and abstracts of the found studies in order to obtain information relating to the studied topic, (3) as result we will obtain candidate studies for each of the selected databases. (4) Evaluate candidate studies based on reading full text and (5) obtain primary studies (selected).

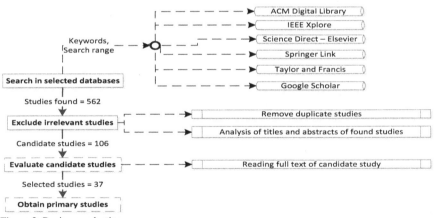

Figure 2: Review conduction

4 Report Review

This section represents the results and the discussion of this review in order to answer the research questions.

4.1 Data Synthesis

Table 1 displays the number of studies retrieved through keyword search in different selected databases. The second column shows the result of the initial selection of items found in each database source selected. The third column shows candidates items who are the numbers of the items that were selected after the exclusion of irrelevant studies. The fourth column shows the selected articles that represents the number of items selected after evaluating them based on a full reading of the text.

Table 1: Number of retrieved and selected studies

Source	Studies found	Candidates	Selected
ACM Digital Library	41	10	2
IEEE Xplore	21	21	5
ScienceDirect-Elsevier	6	4	4
Springer Link	13	6	1
Taylor and Francis	5	0	0
Google Scholar	476	65	25
Total	**562**	**106**	**37**

4.2 Results

It seeks to answer the following questions:

- *What factors or limitations affect the precision of the use case points method estimation?*

To respond to this question, has been made an analysis of all the articles selected and has been identified the factors that affect its precision, which are displayed in the Table 2.

Table 2: Factors affecting accuracy of use case points method

Factors	Articles	(%)	Authors
Environmental factors	11	13,92	[10,19,20,21,22,23,24,25,26,27,28]
Evaluation of use cases complexity	10	12,66	[1,6,17,27,29,30,31,32,33,34]
Lack of standardization of use cases	8	10,13	[23,29,33,34,35,36,37,38]
Technical factors	7	8,86	[10,20,21,23,24,27,28]
Counting number of transactions	7	8,86	[1,17,28,34,37,39,40]
Granularity of use cases	6	7,59	[17,33,36,38,45,48]
Linear relationship between the size and effort	6	7,59	[28,41,42,46,47,50]
Productivity factor	6	7,59	[19,30,41,42,43,44]

Factors	Articles	(%)	Authors
Classification of Actors	5	6,33	[6,17,27,28,34]
Non-functional requirements in use cases	4	5,06	[9,33,41,50]
Different definition of use case transaction	2	2,53	[1,49]
Lack of standardization of definition parameters	1	1,27	[30]
Requirements volatility	1	1,27	[23]
Including historical data	1	1,27	[43]
Model of applied software development process	1	1,27	[29]
Project type	1	1,27	[23]
Selected software framework	1	1,27	[23]
Not addressing a change in requirements	1	1,27	[10]

- *What are the future work to be carried out to improve the use case points method?*

To respond to this question has been carried out an analysis of all the selected articles and has been identified next tasks that will contribute to the improvement of the use case points method. Those tasks are exposed in Table 3.

Table 3: Future work by publishing year

Year	Next tasks	Author
2010	Validate results of the empirical study against standard use case points	[17]
	Lack of standardization of use cases	[29]
2011	Test with larger sizes projects, greater than 5,000 person-hours	[41]
2013	Environmental and technical factors must be updated	[42]
	Use case points should handle use cases of more than 7 transactions	[42]
	The weights of the use cases should be calibrated	[42]
	Expanding experiment with at least 20 projects	[23]
	Expanding experiment with more than 14 projects	[24]
	Consider the effect of the diverse complexity factors.	[25]
2014	Working with use cases that contain at least 21 transactions	[1]
	Extend and include use cases should be considered when estimating the software size	[1]

4.3 Discussions

This section provides arguments in response to the results obtained in the previous section.

Discussion 1: *About the factors or limitations that affect the accuracy of the use case points method estimation*

In reference to the results in Table 2, has been identified five main factors affecting the accuracy of the estimation of use case points technique, which are environmental factors, evaluation of complexity of use cases, lack of standardization of use cases, technical factors and counting number of transactions.

For the environmental and technical factors, has been established questions that are found in this review. The environmental and technical factors affect the estimation [20], so it requires a re-evaluation [21]. For example, a slight variation in the weight value of the environmental factor will increase dramatically the points of the use cases and, therefore, the total effort of the project [22]. Environmental and technical factors are likely to be influenced by the experience of the estimator [23] and do not present standardization [24]. In studies like [25], [26] and [27] has been made a review of the use case points original method by introducing new environmental and technical factors in order to obtain greater precision in the results. Those highlighted the need to refine the parameters used as environmental and technical factors, that are directly related to estimations that are calculated using the use case points method [28].

Regarding the assessment of the use cases complexity, it is recommended to organize the complexity level of each use case depending on its main stages, transactions and dialogs [29], being well known that the ramification of a use case can vary greatly, depending on which design is used [30]. For example, studies like [31] and [32] indicate that a use case of four transactions has twice the weight of a use case that has three transactions. Study [33] shows that in a typical CRUD (Create, Replace, Update, Delete) use case would be correct to see this as a use case of four scenarios, or as a use case with a single scenario [33]. The studies, [27] and [34] propose a new categorization by adding one category to the already existing ones.

About the lack of standardization, the use cases must have an adequate narrative [35]. Popović and Bojić indicate that the biggest problem is the absence of a standard that precisely defines what is a use case [36]. Thomas and Remon prove that the resulted use case points value can vary considerably when applying counting by transactions or scenarios [37]. Many variations of the use cases style can hinder the measure of the use case complexity [38].

In what concerns counting the number of transactions, has been determined that the effectiveness of the use cases points method depends mainly on the way in which the use cases transactions are counted [39]. Also the studies, like [28], [34] and [37] have specified that the criteria for counting the number of transactions is not yet clear. Ochodek, Nawrocki and Kwarciak indicate that transactions count is equivalent to steps count, which in practice is not always true. The difference between the number of steps and the number of transactions affects the value of the use case points [40].

Based on the contributions already described is made necessary reorganize the environmental and technical factors. However, we also need to get a standardization about description of use cases, this being very important since properly standardize a use case prevents more inaccuracies at the time of transactions counting. Finally, regarding to the evaluation of the complexity of use cases it could been determined that has a great dependency to the transactions counting. Analyzing those different studies, it could be inferred that the factors that serve as input for the application of the use case points method are (1) a good application of use cases that is related with the standardization and granularity of use cases, and (2) a correct transactions counting. The above is a starting point to then calculate: classification of actors, complexity of use cases, evaluation of environmental and technical factors, and productivity factor.

Discussion 2: About the future work to be carried out to improve the use case points method

Based on the results in Table 3, it has been able to identify that most authors refer to continue with the refinement of the use case points method by making use of more projects and making use of use cases with more transactions within their research [1] [17] [23] [24] [41] [42]. Other futures work [1] [25] [29] [42] which proposed a standardization of use cases, that the use cases weights should be calibrated and that environmental and technical factors must be updated; have already been evidenced as a subject of study by different authors.

5 Conclusions

The revised research of different authors highlighted that the use case points method need refining in the estimation for software development projects. Based on the analyzed factors it has been able to determine that the inputs for the use case points application method are a standardization of use cases and then a proper counting of the transaction number. Once this made, the imprecision is lower for those later calculations. Studies about the refinement of the estimates based on the improvement of actor's classification, evaluation of complexity of use cases, calculation of environmental and technical factors or calculation of the productivity factors, are useful, reaching their greatest advantage by giving importance to the standardization of use cases and transactions counting in early stage. And regarding the future work, must be revised other factors that affects estimation and also has to be taken into account software projects that have different effort and size.

References

1. Kashyap, D., Misra, A.K., Shukla, D.: Refining the Use Case Classification for Use Case Point Method for Software Effort Estimation. In: Int. Conf. on Recent Trends in Information, Telecommunication and Computing, ITC, pp. 183--191 (2014)

2. Ayyıldız, T.E., Koçyiğit, A.: An early software effort estimation method based on use cases and conceptual classes. Journal of Software. 9(8), 2169--2173 (2014)

3. Recalde, O. M., Aranda, A. M.: Estimación Basada en Casos de Uso UCP – Use Case Points. Tesis de Maestría, Universidad Politécnica de Madrid, Madrid (2010)

4. Pow-Sang, J.A.: Estudio de Técnicas Basadas en Puntos de Función para la Estimación del Esfuerzo en Proyectos de Software. Revista de investigación de Sistemas e Informática. 1(1), 73--82 (2004)

5. Morgenshtern, O., Raz, T., Dvir, D.: Factors Affecting Duration and Effort Estimation Errors in Software Development Projects. Information and Software Technology. 49 (8), 827--837 (2007)

6. Nassif, A.B., Capretz, L.F., Ho, D.: Software estimation in the early stages of the software life cycle. In: International conference on emerging trends in computer science, communication and information technology, pp. 5--13 (2010)

7. Mishra, S., Pattnaik, P.K., Mall, R.: Early Estimation of Back-End Software Development Effort. International Journal of Computer Applications, 33(2), 6--11 (2011)

8. Wen, J., Li, S., Lin, Z., Hu, Y., Huang, C.: Systematic literature review of machine learning based software development effort estimation models. Information and Software Technology. 54(1), 41--59 (2012)

9. Sharma, A., Kushwaha, D.S.: Applying requirement based complexity for the estimation of software development and testing effort. ACM SIGSOFT Software Engineering Notes, 37(1), 1--11 (2012)

10. Singh, J., Sahoo, B.: UML Based Object Oriented Software Development Effort Estimation Using ANN. National Institute of Technology Rourkela, Rourkela (2012)

11. Subriadi, A.P., Sholiq, Ningrum, P.A.: Critical review of the Effort rate value in use case point method for estimating software development effort. Journal of Theoretical and Applied Information Technology, 59 (3), 735--744 (2014)

12. Saroha, M., Sahu, S.: Analysis of various Software Effort Estimation Techniques. International Research Journal of Computers and Electronics Engineering (IRJCEE), 3(2), 1--7 (2015)

13. Karner, G.: Resource Estimation for Objectory Projects. Objective Systems SFAB (1993)

14. Kusumoto, S., Matukawa, F., Inoue, K., Hanabusa, S., Maegawa, Y.: Effort estimation tool based on use case points method. Osaka University (2005)

15. Clemmons, R.: Project Estimation With Use Case Points. The Journal of Defense Software Engineering, 18--22 (2006)

16. Wang, F., Yang, X., Zhu, X., Chen, L.: Extended use case points method for software cost estimation. In: International Conference on Computational Intelligence and Software Engineering, CiSE 2009, pp. 1--5, IEEE, China (2009)

17. Nunes, N.J.: iUCP-estimating interaction design projects with enhanced use case points. In: England, D., Palanque, P., Vanderdonckt, J., Wild, P. (eds.) Task Models and Diagrams for User Interface Design, vol. 5963, pp. 131--145. Springer, Berlin/Heidelberg (2010)

18. Satapathy, S.M., Rath, S.K.: Use case point approach based software effort estimation using various support vector regression kernel methods. International Journal of Information Processing, 7(4), 87--101 (2014)

19. Azzeh, M., Nassif, A. B.: A hybrid model for estimating software project effort from Use Case Points. Applied Soft Computing (2016)

20. Habib, M. U., Ali, M. A., Atique, N.: Extending the UCP Model by Incorporating the Prevailing Trends in Software Effort Estimation. International Journal of Computer Applications, 59(5), 1--7 (2012)

21. Ribeiro, B. B., Modesto, D. M., Santos, G.: Avaliação da Importância dos Fatores Técnicos e Ambientais do Método Pontos por Caso de Uso com Base no Método AHP. In: CIbSE, pp. 210--223 (2012)

22. Anukula, J.M., Perumal S.M.: Analog-Based Cost Estimation For Managing Inconsistency In Software Development. International Journal of Research Sciences and Advanced Engineering, 3(2), 50--54 (2012)

23. Alves, R., Valente, P., Nunes, N. J.: Improving software effort estimation with human-centric models: a comparison of UCP and iUCP accuracy. In: Proceedings of the 5th ACM SIGCHI symposium on Engineering interactive computing systems, pp. 287--296, ACM, New York (2013)

24. Ayyıldız, T.E., Koçyiğit, A., Peker D.: Comparison of Three Software Effort Estimation Methodologies with Case Study. In: 3rd World Conference on Innovation and Computer Science, 4, pp. 257--262 (2013)

25. Azzeh, M.: Software cost estimation based on use case points for global software development. In: Computer Science and Information Technology (CSIT), 2013 5th International Conference on, pp. 214--218, IEEE, Ammán (2013)

26. Jha, P., Jena, P.P., Malu, R.K.: Estimating Software Development Effort using UML Use Case Point (UCP) Method with a Modified set of Environmental Factors. International Journal of Computer Science and Information Technologies, 5(3), 2742--2744 (2014)

27. Kirmani, M. M., Wahid, A.: Impact of Modification Made in Re-UCP on Software Effort Estimation. Journal of Software Engineering and Applications, 8(6), 276--289 (2015)

28. Kirmani, M. M., Wahid, A.: Use Case Point Method of Software Effort Estimation: A Review. International Journal of Computer Applications, 116(15), 43--47 (2015)

29. Remón, C. A., Thomas, P. J.: Análisis de Estimación de Esfuerzo aplicando Puntos de Caso de Uso. In: XVI Congreso Argentino de Ciencias de la Computación (2010)

30. Felipe, N. F., et al.: A Comparative Study of Three Test Effort Estimation Methods. Revista Cubana de Ciencias Informáticas, 8, 1--13 (2014)

31. Nassif, A.B., Capretz, L.F., Ho, D.: Enhancing use case points estimation method using soft computing techniques. Journal of Global Research in Computer Science, 1(4), 12--21 (2010)

32. Nassif, A.B., Capretz, L.F., Ho, D.: Calibrating use case points. In: Companion Proceedings of the 36th International Conference on Software Engineering, pp. 612--613, ACM, New York (2014)

33. Kamal, M.W., Ahmed, M.A.: A proposed framework for use case based effort estimation using fuzzy logic: building upon the outcomes of a systematic literature review. International Journal of New Computer Architectures and their Applications (IJNCAA), 1(4), 953--976 (2011)

34. Park, B.K., Moon, S.Y., Kim, R.Y.C.: Improving Use Case Point (UCP) Based on Function Point (FP) Mechanism. In: 2016 International Conference on Platform Technology and Service (PlatCon), pp. 1--5, IEEE, Korea (2016)

35. Bajaj, P., Bathla, D.R.: A Tool to Evaluate the Performance of UCP. International Journal of Advance Research in Computer Science and Management Studies, 2(7), 84--89 (2014)

36. Popović, J., Bojić, D.: A comparative evaluation of effort estimation methods in the software life cycle. Computer Science and Information Systems, 9(1), 455--484 (2012)

37. Thomas, P.J., Remón, C.A.: Análisis comparativo de estimación de esfuerzo en el desarrollo de software. In: XVII Congreso Argentino de Ciencias de la Computación, Mar del Plata (2011).

38. Chaudhary, A., Chaudhary, N., Khatoon, A.: Analysis of Use Cases and Use Case Estimation. International Journal Of Engineering And Computer Science, 4(3), 10791--10798 (2015)

39. Jena, P.P., Mishra, S.: Survey Report on Software Cost Estimation Using Use Case Point Method. International Journal of Computer Science & Engineering Technology, 5(4), 280—287 (2014)

40. Ochodek, M., Nawrocki, J., Kwarciak, K.: Simplifying effort estimation based on Use Case Points. Information and Software Technology, 53(3), 200--213 (2011)

41. Nassif, A.B., Capretz, L.F., Ho, D.: A Regression Model with Mamdani Fuzzy Inference System for Early Software Effort Estimation Based on Use Case Diagrams. In: Third International Conference on Intelligent Computing and Intelligent Systems, pp. 615--620 (2011)

42. Nassif, A.B., Ho, D., Capretz, L.F.: Towards an early software estimation using log-linear regression and a multilayer perceptron model. Journal of Systems and Software, 86(1), 144--160 (2013)

43. Silhavy, R., Silhavy, P., Prokopova, Z.: Algorithmic Optimisation Method for Improving Use Case Points Estimation. PloS one, 10(11), 1--14 (2015)

44. Azzeh, M., Nassif, A.B., Banitaan, S.: An Application of Classification and Class Decomposition to Use Case Point Estimation Method. In: 2015 IEEE 14th International Conference on Machine Learning and Applications (ICMLA), pp. 1268--1271, IEEE, Miami (2015)

45. Heo, R., Seo, Y.D., Baik, D.K.: An Elementary-Function-Based Refinement Method for Use Cases to Improve Reliability of Use Case Points. Journal of KIISE, 42(9), 1117--1123 (2015)

46. Alves, L.M., Sousa, A., Ribeiro, P., Machado, R.J.: An Empirical Study on the Estimation of Software Development Effort with Use Case Points. In: 2013 IEEE Frontiers in Education Conference (FIE), pp. 101--107, IEEE, Oklahoma (2013)

47. Popovic, J., Bojic, D., Korolija, N.: Analysis of task effort estimation accuracy based on use case point size. IET Software, 9(6), 166--173 (2015)

48. Bone, M.A., Cloutier, R.: Applying Systems Engineering Modeling Language (SysML) to System Effort Estimation Utilizing Use Case Points. INCOSE International Symposium, 21(1), 14--127 (2011)

49. Ochodek, M., Alchimowicz, B., Jurkiewicz, J., Nawrocki, J.: Improving the reliability of transaction identification in use cases. Information and Software Technology, 53(8), 885--897 (2011)

50. Nassif, A.B., Ho, D., Capretz, L.F.: Regression Model for Software Effort Estimation Based on the Use Case Point Method. In: 2011 International Conference on Computer and Software Modeling, 14, pp. 106--110 (2011)

Part II
Knowledge Management

Teaching Computer Programing as Knowledge Transfer: Some Impacts on Software Engineering Productivity

Orlando López-Cruz[1], Alejandro León Mora[1], Mauricio Sandoval-Parra[1], Diana Lizeth Espejo-Gavilán[1],

[1] Universidad El Bosque, Av. 9 132-A-01 Bloque M. Piso 3, Bogotá D.C. 110121, Colombia
{orlandolopez, alejandroleon, mauriciosandoval, despejo}@unbosque.edu.co

Abstract. Programming skills of software engineers that affect software development productivity are central to any of the computing disciplines. While literature focuses on how to teach novice programmers, the aim of this research is to show how to strengthen programming skills of programmers by effectively transferring knowledge to those who had bad experiences when learning computer programming or have not developed enough programming skills to get a productivity standard. Since software engineering is a knowledge-intensive application discipline, a knowledge transfer process is conducted to improve the productivity of computer programmers involved in software engineering projects. An *ad-hoc* methodology allowed to follow-up changes that revealed that improvements in the capability to absorb new external knowledge increases overall productivity of individuals in software development teams. This finding may be useful for software companies looking for increasing their productivity.

Keywords: Knowledge Transfer, Knowledge Management, Teaching Computer Programming, Software Engineering.

1 Introduction

Computing disciplines have been identified as Computer Engineering, Information Technology, Information Systems, Software Engineering and Computer Science [1] but neither students nor businesses differentiate them [2]. Even worse, in some countries, academic programs do not hold any of those names but 'systems engineering' which, according to INCOSE denotes "*an engineering discipline whose responsibility is creating and executing an interdisciplinary process to ensure that the customer and stakeholder's needs are satisfied in a high quality, trustworthy, cost efficient and schedule compliant manner throughout a system's entire life cycle*" [3], this is to say that "systems engineering" refers to a broader body of knowledge than just software engineering, computer programming, or any of the computing

© Springer International Publishing AG 2017   145
J. Mejia et al. (eds.), *Trends and Applications in Software Engineering*, Advances
in Intelligent Systems and Computing 537, DOI 10.1007/978-3-319-48523-2_14

disciplines. However, any of these disciplines require to develop computer programming skills, especially when people is going to be involved in software development projects.

Many studies spend efforts to make distinctions between these disciplines, but this research focuses on something that they share in common. They share the need to produce computer programs (i.e. software) whether as an art [4-6] or science [7] or, more plausible, as engineering [8]. The issue tackled in this study is not how to make the software seem to do what is supposed to do, but how software can be produced minimizing programmers time. This research draws the attention to the fact that behind computer program development there are computer programmers. People that is responsible to produce software from the requirements phase to the operations and maintenance phase. People in need of training on software development methodologies. People in need of interacting with others in respectful ways in a team work.

Many other studies have been conducted in order to overcome difficulties involved in teaching programming in an introductory course [9, 10] but, to the best of our knowledge, no studies report on results enforcing computer programming abilities in programmers (coming from any computing discipline) involved in a project. A question arose at this point: What is the impact of computer programming knowledge transfer on software productivity? This led to involve senior students of a "*systems engineering*" program in a real software development project, working at a client place next to experienced developers. Individuals were immersed in a stress-controlled but real software development and implementation environment.

The aim of this paper is to show results of a research focused in increasing productivity of programmers by improving their programming skills from a knowledge-based centered process. In addition, this is new because studies are centered to engage freshmen students in programming disciplines but not to retain workforce and improve their skills.

This paper is structured as follows. First, the phases defined to guide the research, then a succinct section of the relationship between the concepts of knowledge management and software is introduced. Then, the sections of results analysis and conclusions are developed.

2 Methodological Issues

In order to conduct a research regarding the real world either to explain it or transform it, qualitative and quantitative approaches are sides of the same coin: both, a qualitative approach to meanings and a quantitative approach to facts are needed [11]. In fact, The distinction between qualitative and quantitative research methodologies is ideological [12]. This is to say that complex phenomena, such as software development, should be studied both in its quantitative dimensions, and its qualitative dimensions as well. Even more, that an emphasis on quantitative or qualitative issues is not a priority. A different enriched methodological approach must be addressed when dealing with complex environments [13, 14] or phenomena such as the one being reported in this paper.

2.1 The Phases of the Research

The methodology is structured in four phases. (i) Preparation and selection phase consists of two parallel steps: first, determination of the environment to conduct the experiment. The right environment is a real software development project that may provide space to conduct tutoring activities. And, second, recruiting of senior students of an undergraduate program of a computing discipline from those with the lowest grades in computer programming courses. An assessment of students´ abilities is conducted to state a baseline of knowledge. Candidates are interviewed to validate their negative attitude towards computer programming. From this interviews arise assessment categories (Table 1). These categories are more specific than those presented in other studies [15].

(ii) During immersion phase, selected individuals are exposed to a real project that is being conducted. They are informed about the problem to be solved by the product resulting from software development. Tutoring meetings are scheduled in order to improve their programming abilities.

Table 1. Categories to assess knowledge internalization of each subject of study.

Id	Category	What to assess
C1	Interest to learn computer programming.	Basic interest to learn to program computers by himself or by means of the support of a tutor.
C2	Computer programming language knowledge.	The level of understanding and usage of the Java® programming language.
C3	Teamwork	Previous experience working with unknown people in team activities.
C4	Project control	Compliance with assignments, advances on assignments, deadlines, and deliverables (as well in programming code as in documentation).
C5	Computer technology	Specific knowledge of computer technology to use during project development both in software development and communications and collaborative software supporting development activities.
C6	Methodology	Knowledge and usage of the methodology used in the software development project.
C7	Frameworks learning	Understanding of *PrimeFaces* and *Spring* frameworks. Usage of this frameworks in the project.
C8	Database modeling and database management from code.	Learning to manage the database from Java code.
C9	Object Oriented Programming	Knowledge of the object-oriented programming paradigm in practice.
C10	Usability basic practices	Analysis, learning and usage of basic usability practices to produce a functional and easy to use software application.

(iii) The third phase is a hands-on learning process and follow-up. Each individual is instructed to complete a field diary. By means of field diaries, monitoring and guidance are provided. Advances are followed-up.

(iv) The fourth Phase consists of the final assessment procedure to compare actual abilities and knowledge that each individual actually exhibits with respect to the knowledge baseline that was recorded in phase one.

These phases conform a scene to induce improvement in knowledge for each individual involved in this study, as well as a general stage for an ongoing assessment process.

2.2 Issues in Conducting the Phases of This Research

When in phase one, besides looking for an actual software development project, an environment allowing effective tutoring to individuals is required. In addition, this should not disturb actual software development and avoid client concern or annoyance. This was not easy to set. However, the research was conducted in a real scenario where the client was asking to develop a software piece to capture data on medical variables from biomedical devices, including manual data gathered when using devices such as tensiometers and stethoscopes at a pediatric intensive care unit. The assessment categories introduced in Table 1 were refined by reviewing personal logs of selected students.

In order to recruit volunteers, students from a senior cohort were selected according to their low performance (low grades) in computer programming courses. In addition, those that were selected were asked to express their opinion in relation to programming in practice. Those who expressed dissatisfaction toward computer programming, low level of knowledge of programming languages and programming paradigms were preferred. Interest to learn and their abilities to work as team members were determined by means of questionnaires. This allowed to identify the initial programming abilities of the individuals and to set a baseline. Finally, some of them were selected. Follow-up was done by inspection of development of user stories and field diaries, and recording and tackling difficulties arising in the ongoing project

This paper reports results from two of the volunteer students that were involved in the experimental process. They never worked before as a part of a software development team. They received user stories that were refined by group meetings including members of the development team and final users from the client organization, by using activity (hands-on experience) records and personal logs (one for each individual). This last element was crucial to assess changes occurred in individuals, especially when comparing to the baseline assessment.

The process was divided into "learning stages" and "stages for practice". During learning stages, individuals were asked to read chapters of different books on object oriented programming, or contents of web pages, in order to complement or refresh programming concepts. Meetings were held to provide support regarding some topics required to proceed to develop. Stages of practice were guided by IEEE 1074 tailoring the software development life cycle [16]. At the first (practice) stage, individuals showed low performance. This situation led to a delay in user stories deployment. The researchers were not disappointed with this gross result, as this was supposed to happen. Patience was definitely worthwhile. After individuals involved progressively with the project and feeling confident with themselves, a work team was consolidated.

However, in spite of the fact that individuals devoted about 25 hours per week to produce code and documentation deliverables of the project, for practice and learning stages, it took longer to accomplish. However, results encouraged individuals to improve compliance with deadlines to the point they accomplished to deploy user stories as soon as the client were requiring them. This enthusiastically encouraged individuals to look for the development of deliverables in order to get user satisfaction.

A category, "interest to learn" (Table 1) was central to the development of the classification from individual logs reviews, deliverables from user stories summing up thirty-one modules [17]. The overall involvement of the subjects in the project was divided into four stages, identified as stages 1 to 4, each one of a three months period. Each stage was followed immediately by another.

During the first stage, subjects were trained and self-trained on the computer technologies required to be engaged in the project. This allows individuals to adapt daily to the work of software development.

3 Knowledge Transfer and Software

Recent attention has been paid to knowledge transfer in software engineering [15] either understanding knowledge as a main asset in software organizations [18 p.26, 19 p. 105] or because it is a knowledge intensive discipline [18, 20] as well as a computer programming skill [9] demanding activity. Even so, the software industry has been recognized as an "engineering" endeavor but of a different kind [8]. The reason is that software is manufactured once (then deployed many times) and it is essentially an abstract product, or at least with no physical component. Software as a product is more like a book of poems than a bridge. Both are produced once, but the physical dimension of a bridge is necessary for its usage while the physical part of a book (paper, ink and so on) may be abstracted, for instance, by publishing it as an electronic book. In addition, this implies that statistical quality control may not be applied to the software production process [21]. The essence of the book is the knowledge that has been codified: the poem. Even better, the codification of knowledge is what makes the essence of the book. The same applies to software, even when it is maintained [22]. Nonetheless, up to this point, there is nothing completely new.

What is new is to focus this research on the 'workforce' to produce software. Software development is a creative process that is conducted by human beings at their intellectual level. In this context, software engineers (or computer scientists, or "systems engineers") and poets or writers are alike. Their challenge is to conduct intellectual processes to produce a result that is a unique instance of a class of abstract objects.

4 Results Analysis

In order to respond the question asked in the introduction, assessment records were plotted on a graph (Fig. 1). Both subjects (subject 1 and subject 2) were exposed to the same project in four stages (listed from 1 to 4 horizontally in Fig. 1 for subject 1, and in Fig.2 for subject 2).

Since the research is focused in what is changed in a specific environment (i.e. the interest is focused more on the ongoing process than on final stand-alone results), the changes between stages were observed. Therefore, for each stage 1 to 4, the same categories C1 to C10 are assessed. Each category was graded from 1 to 10 (arrayed vertically). Grades from 1 to 3 were considered Low, 4 to 7 Medium, and 8 to 10 High.

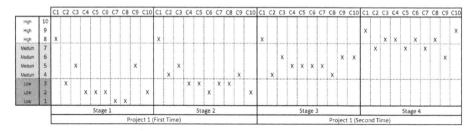

Fig. 1. Knowledge transfer assessment for subject A. Horizontally, four stages are depicted (*stages 1 to 4*). During stages 1 to 4 each of the ten categories in Table 1 were assessed.

Subject 1 in stage 1 got just one item High (C1), two items Medium (C3 and C9), and the remaining items were graded Low (Fig. 1). During stage 2 the item C1 continued High, while item C2 passed from Low to Medium. C3 and C9 stayed Medium with a little local decrease of C9 from 5 to 4. While the items C4,C5,C6,C7,C8, and C10 remained Low, it was encouraging the local change of its grading from 'very low' values to values higher in the same interval.

Stage 3 for subject 1 was a qualitative jump in knowledge categories assessment from mainly Lows to mainly Mediums. And stage 4 led to an unexpected mainly Highs and upper Mediums (Fig.1). Just C3, C5, C7, and C9 of subject 1 remained Medium. Examining each of the items, C1: Teamwork, C5: Computer Technologies, C7: Frameworks learning, and C9: Object Oriented Programming were the items in upper Medium.

For the case of subject 2 (Fig. 2), assessment of the categories C1 to C10 during stage 1was not qualitatively different from subject 1. This means that grades for categories being assessed were mainly Low. Just C3, C9, and C10 were in the Medium Interval. However, C9 and C10 were at the lowest Medium grade.

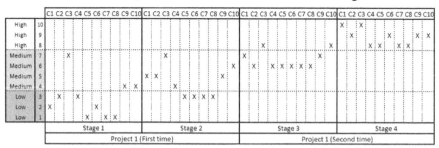

Fig. 2. Knowledge transfer assessment for subject B. Horizontally, four stages are depicted (stages 1 to 4). During stages 1 to 4 each of the ten categories in Table 1 were assessed.

During stage 2, there was a significant improvement in grades in the overall set of categories. They were assessed mainly in the Medium interval and just C5, C6, C7, and C8 were at the Low interval but at the upper grade of the interval.

Stage 3 proved to be an important improvement for subject 2. Must of the categories were at the upper Medium interval, and C3 and C10 were at the lower High interval. But stage 4 revealed outstanding results for this subject. Every category was graded at the High interval. Just four out of ten categories, C4, C5.C7, and C8 were at the lowest grade of the High interval.

The lowest graded categories at stage 4 for subject 2 were C4: Project control, C5 Computer Technology, C7: Frameworks learning, and C8: Database modeling and database management from code. When comparing this list with the lowest categories at stage 4 for subject 1, C5 and C7 are in common.

In addition, it may be observed that the category C3: teamwork was not fully developed, that may be explained by a low slope during stages 1 and 2, which in turn could be explained by externalities. From stage 1 to stage 3 the increase was not significant, but from stage 3 to stage 4 an unexpected and relatively significant increase was observed in the overall set of categories being assessed.

In order to check the consistency of this assessment, an additional measurement was considered: the user stories deliverables. These were increasing from stage to stage also, which means that subjects achieved higher levels of productivity as time goes by.

Subjects selected for this research exhibited poor or limited computer programming skills. Both subjects reported in this study continued during the first two stages of the experimental process to display real difficulties on a range of fundamental skills for integrating to a software development project, not just because of their low programming skills but because of their low profile in abilities like teamwork and project control. This is because the simple model of knowledge transfer [23] consisting of agent A making knowledge available to the environment of Agent B, as it were the classical data communication model [24], does not reveal the essence of knowledge transfer. Knowledge transfer is not a matter of data communication as in information theory. For knowledge to be effectively transferred, the receiver (Agent B) must not be a passive agent, but must exhibit the dynamic capability to absorb the knowledge [25, 26] available to make it productive.

The results shown in stages 3 and 4 support the statement that absorptive capacity of individuals or organizations [25, 26] must be developed before knowledge may be exploited by the receiver. In this context, the capability to increase the number of deliverables by the subjects of study involved in a software development project.

5 Conclusion

This paper has introduced a hands-on experiment to teach computer programming while "learners" are involved in a "true" software development project. It is worth noting that while the current interest of many researchers is focused in novice programmers, this experiment was conducted over individuals of a computing discipline expressing negative experiences towards computer programming with low-level computer programming skills. This is to say, that the focus is to find practical ways to improve skills –capabilities of agents- to "produce" software deliverables of the project, not just coding programs and ensuring their correctness [7]. Or to phrase it another way, this research is focused to improve individual skills from those who has previous knowledge of computer programming and, however, despite of it, they have not reached some productivity standards.

The experiment conducted was not a set of training sessions or the development of an educational course syllabus. The experiment was a knowledge transfer process. What was ensured was the process of developing the capabilities of the receiving individuals to make computer programming knowledge productive in a real environment. The results observed on the individuals under study allow infer that productivity was significantly increased in a relatively short period of time, as a result of a controlled process of knowledge transfer.

From the experiment that was conducted, it was found that the ten categories (Table 1) that defined the set of assessment parameters showed that in subjects under study reveal a knowledge absorption process and knowledge seizing by exploiting developed (and developing at the individual level) programming skills in a real software engineering project environment. Individuals were involved on a part-time basis in this study. It could be thought that on a full-time basis an improvement in software productivity may be achieved in a shorter time.

A generalization of this finding is still an issue because knowledge is not a matter of data accumulation, but a cognitive process. Knowledge transfer could not be measured directly, so proxy variables such as those of the categories in Table 1 were measured to obtain an indirect estimate of the knowledge effectively transferred. This opens an opportunity to conduct research in working teams of real software engineering projects about productivity increase by improving absorptive capacities regarding specific categories in a similar way as the categories (Table 1) involved in this study.

Acknowledgments. Authors express their gratitude to Hospital Santa Clara, Bogotá D.C. for authorizing access to their software project at the Pediatric Intensive Care Unit. Especial thanks to Dr. Armando León Villanueva, Pediatric Lung Care and Pediatric Pulmonologist of the Pediatric Intensive Care Unit, and Dr. Maria Claudia Guzmán Diaz Pediatrician of Universidad El Bosque and 'Hospital Cardiovascular del Niño' San Mateo, Cundinamarca. Authors would further like to thank three anonymous reviewers for their useful comments and feedback which have helped to write the final version of this paper.

References

1.   ACM-IEEE, *Computer Science Curricula 2013: Curriculum Guidelines for Undergraduate Degree Programs in Computer Science - December 20, 2013. ISBN: 978-1-4503-2309-3.* 2013, United States of America: The Joint Task Force on Computing Curricula Association for Computing Machinery (ACM) IEEE Computer Society.

2.   Courte, J. and C. Bishop-Clark, *Do students differentiate between computing disciplines?* ACM SIGCSE Bulletin. **41**(1): p. 29-33. (2009)
3.   INCOSE. *What is Systems Engineering. The International Council on Systems Engineering.* http://www.incose.org/AboutSE/WhatIsSE.
4.   Knuth, D.E., *The art of computer programming: Sorting and searching.* Vol. 3, Reading, Massachussets: Addison-Wesley. 426-458. (1999)
5.   Knuth, D.E., *The art of computer programming Fundamental Algorithms.* Vol. 1, Reading, Massachussets: Addison-Wesley. World students series. (1973)
6.   Knuth, D.E., *The art of computer programming: Seminumerical algorithms.* Vol. 2, Reading Massachussets: Addison-Wesley (1973).
7.   Gries, D., *The science of programming.* Springer Science & Business Media. (2012)
8.   Bryant, A. *It's engineering Jim... but not as we know it: software engineering—solution to the software crisis, or part of the problem?* in *Proceedings of the 22nd international conference on Software engineering.* 2000. ACM. (2000)
9.   Rubiano, S.M.M., O. López-Cruz, and E.G. Soto. *Teaching computer programming: Practices, difficulties and opportunities.* in *Frontiers in Education Conference (FIE), 2015. 32614 2015. IEEE.* (2015)
10. Plonka, L., et al., *Knowledge transfer in pair programming: An in-depth analysis.* International journal of human-computer studies. **73**: p. 66-78. (2015)
11. González López, J.L. and P. Ruiz Hernández, *Investigación cualitativa versus cuantitativa:¿ dicotomía metodológica o ideológica?* Index de Enfermería. **20**(3): p. 189-193.(2011)
12. Monzón Laurencio, L.A., *Ni cualitativo ni cuantitativo: un estudio hermenéutico analógico sobre la metodología de la investigación.* (2011)
13. Mejía, A., et al. *Ser directo puede traerte problemas, pero ser indirecto también: Las realimentaciones en dinámica de sistemas cualitativa y cuantitativa.* in *Artículo aceptado para el Congreso Latinoamericano de Dinámica de Sistemas.* (2007)
14. Aceros, V., et al., *¿Qualitative or quantitative? That's not the question: a method for developing dynamic hypotheses.* , in *Proceedings of the 9th Latin American System Dynamics Conference.* Universidade de Brasília: Brasilia. (2011)
15. Camacho, J.J., J.M. Sanches-Torres, and E. Galvis-Lista, *Understanding the Process of Knowledge Transfer in Software Engineering: A Systematic Literature Review*, in *The International Journal of Soft Computing and Software Engineering [JSCSE]. Special Issue: The Proceeding of International Conference on Soft Computing and Software Engineering 2013 [SCSE'13], Doi: 10.7321/jscse.v3.n3.33 e-ISSN: 2251-7545.* 2013: San Francisco State University, CA, U.S.A. p. 219-229. (2013)
16. Fitzgerald, B., N. Russo, and T. O'Kane, *An empirical study of system development method tailoring in practice.* ECIS 2000 Proceedings, p. 4 (2000)
17. Cresswell, J.W., *Research design Qualitative, quantitative and mixed methos approaches.* Sage Publications. (2009)
18. Rus, I. and M. Lindvall, *Knowledge management in software engineering.* IEEE software. **19**(3): p. 26. (2002)
19. Mathiassen, L. and P. Pourkomeylian, *Managing knowledge in a software organization.* Journal of Knowledge Management. **7**(2): p. 63-80. (2003)
20. Ward, J. and A. Aurum, *Knowledge management in software engineering-describing the process*, in *Australian Software Engineering Conference, 2004. Proceedings. 2004.* IEEE. p. 137-146. (2004)
21. Basili, V.R. and G. Caldiera, *Improve software quality by reusing knowledge and experience.* MIT Sloan Management Review. **37**(1): p. 55. (1995)
22. Batista Dias, M.G., N. Anquetil, and K.M. de Oliveira, *Organizing the knowledge used in software maintenance.* J. UCS. **9**(7): p. 641-658. (2003)

23.   Ajith Kumar, J. and L. Ganesh, *Research on knowledge transfer in organizations: a morphology.* Journal of knowledge management. **13**(4): p. 161-174. (2009)
24.   Shannon, C.E., *A mathematical theory of communication.* ACM SIGMOBILE Mobile Computing and Communications Review. **5**(1): p. 3-55. (2001)
25.   Lopez-Cruz, O. and N. Obregon Neira, *Diseño de la capacidad de absorción en las organizaciones: propuesta de un nuevo constructo y literatura*, in *Congreso Nacional e Internacional en Innovación en la Gestión de Organizaciones, Abril, 2016*, F.R. Santoyo, Editor. Universidad Central: Bogotá. p. 222-237.(2016)
26.   López-Cruz, O. and N. Obregón Neira, *Design of the Organizational Absorptive Capacity: A New Construct Proposal and Literatures.* In publishing, (2016)

Improving Competitiveness Aligning Knowledge and Talent Management with Strategic Goals

Arturo Mora-Soto, Cuauhtemoc Lemus-Olalde, Carlos A. Carballo

Centro de Investigación en Matemáticas, A.C.
Avenida Universidad, 222, 98068, Zacatecas, México
{jmora, clemola, abraham}@cimat.mx

Abstract. In recent years, knowledge management has attracted the attention of the software engineering community. The ability of a software development organization to manage the knowledge that resides in its employees provides it the opportunity to improve its productivity, up to the point of obtaining a sustainable competitive advantage over other organizations. However, achieving a successful implementation of a knowledge management initiative is not trivial. Sometimes when a knowledge management initiative is addressed in a software engineering organization the main error that is committed is to focus most of the efforts in defining a technological solution ignoring human aspects of great importance as talent. This paper proposes a framework intended to allow software engineering organizations to develop and implement knowledge and talent management strategies in a comprehensive and systematic way in order to help these organizations to achieve their business goals. The framework proposed by authors is illustrated using a real case study where it was used in a software development organization to improve work-teams' productivity.

Keywords: Knowledge management; software engineering; strategic goals; talent management.

1. Introduction

Software engineering is a knowledge-intensive activity [1]–[5], and the most valuable asset for software engineering organizations is the knowledge and talent that resides in their members' minds [1], [5]. For the purpose of this work authors understand **knowledge** as "*a fluid mix of framed experience, values, contextual information, and expert insight that provides a framework for evaluating and incorporating new experiences and information. In organizations, it often becomes embedded not only in documents or repositories but also in organizational routines, process, practices, and norms* [6]"; while **talent** represents the individual flairs that give value to people and is the driving force behind an individual's job performance, it is defined as "*a recurring pattern of thought, feeling, or behavior that can be productively applied* [7]". Knowledge could be made explicit for future reuse, but talent can't be encapsulated or taught, and it represents the side of the human intellect that only can be used on the spot.

© Springer International Publishing AG 2017   155
J. Mejia et al. (eds.), *Trends and Applications in Software Engineering*, Advances in Intelligent Systems and Computing 537, DOI 10.1007/978-3-319-48523-2_15

The importance and benefits of managing knowledge and talent in organizations has been long recognized in the literature [6], [8]. In recent years, the software engineering community has shown great interest in this field, and has recognized the need of managing knowledge and talent due to the benefits that can bring to software engineering organizations [1], [2], [4], [5], [9]–[11] such as improving the productivity of software engineering organizations, improving software development process, decreasing time and costs of development, increasing software product quality, or making better decisions, in sum improving their competitiveness. However it is necessary to consider that achieving a successful knowledge and talent management implementation is not a simple task, there are many challenges and obstacles to overcome [1].

Many studies have identified different errors committed in the way that knowledge and talent management initiatives are addressed by organizations [1], [3], [6], [10], [12]–[15]; these studies present different recommendations and factors that should be taken into account to successfully manage the organizational knowledge and talent [1], [3], [5], [6], [10], [12]–[23].

One of **the most important recommendations is the definition of a knowledge and talent management strategy** [3], [19], [20]; this strategy should be associated with the organizational goals [15], [21]–[23], it must take into account every organizational and technological factor that affects knowledge management [3], [10], it must follow a systematic approach [10], and finally, it must allow to be continually evaluated and fed back [3], [20].

The lack of a knowledge and talent management strategy is a common factor found in different software engineering organizations that have failed in their knowledge and talent management initiatives, or have some kind of deficiency in such initiatives [10], [24], [25].

While there has been a growing interest in studying and implementing knowledge and talent management strategies in software engineering organizations, most of the work done is focused on information technologies. More attention must be paid to the knowledge and talent management as a dimension of the strategy of an organization, this means that organizations must develop a strategic approach in order to manage their organizational knowledge and talent, which aims to build, nurture, and fully exploit knowledge and talent across systems, processes, and people [5].

Considering the importance of managing knowledge and talent in a software engineering organization to improve its competitiveness; the question that arises is: How can the organization develop and implement a knowledge and talent management strategy, bearing in mind all the edges that the problem of achieving a successful implementation has?

Authors proposal is a framework that allows software engineering organizations to develop and deploy knowledge and talent management strategies, accordingly with organizations' reality and needs, in a comprehensive and systematic way, and that can be continuously evaluated and feedback.

2. Proposal: knowledge and talent management framework for software engineering organizations

2.1. Proposal Overview

The concept of *knowledge and talent management strategy* (*KT-MS*) is used with different meanings [23], however none of them fit in the purpose of authors' proposal. Due to this fact, and in order to avoid misunderstandings, authors define their knowledge and talent management strategy as *a set of ten elements that are adaptable to the organization's nature and constraints, and defines how a software engineering organization must manage its knowledge and talent in order to achieve its organizational goals.* The proposed framework has two main components:

- **Knowledge and talent management strategy.** This strategy describes ten elements that need to be developed in the organization and their characteristics; these elements are called artifacts of the KT-MS.
- **Phases and processes.** This component defines how the organization must develop and deploy each of the artifacts of the strategy.

2.2. General Requirements

This framework has been designed based on a set of general requirements defined to avoid the problems found in the literature and ratified in practice. These requirements set up that a *KT-MS* must meet the following characteristics.

- **To be aligned with the business goals**, otherwise the knowledge and talent management is meaningless.
- **To consider organizational and technological factors affecting knowledge and talent management in a systematic way**, if not, its complexity makes them difficult to handle.
- **To fit the reality of the organization** in order to define an appropriate KT-MS for each organization. Due to the complexity of knowledge and talent management and the complexity of organizations, it is virtually impossible to have standard solutions for managing knowledge and talent; however, linking the KT-MS with organization reality could be a key success factor.
- **To evolve according to the organization needs.** Both, knowledge and organization evolve continuously, so knowledge and talent management must evolve with them too.
- **To be deployed using the technology that best fit the organization's culture and human resources traits.** This implies the incorporation of professionals from social disciplines to help in the guidelines deployment. Due to the different nature of talent and knowledge, organizations need specific technologies to manage each.
- **To represent organizational knowledge**, in order to encode, store, and retrieve the knowledge that is generated into the organization, to make visible the talent of the organization members, and leverage knowledge and talent towards the organization's business goals.

2.3. Knowledge and Talent Management Strategy

Authors proposed that a *KT-MS* must be comprised by the following ten elements; these elements are called *artifacts of a knowledge and talent management strategy*.

- *Objectives*. Define what the software engineering organization hopes to achieve through knowledge and talent management. These must be properly aligned with the business goals.
- *Scope*. Determines what areas, projects or products of the software engineering organization will be involved, these are called members of the scope.
- *Approach*. Determines for each of the members of the scope, if the knowledge will be mainly managed through codification, for explicit knowledge, or personalization, for tacit knowledge.
- *Knowledge assets*. Represent a set of knowledge to be managed by the organization, it could be formed by explicit knowledge and tacit knowledge. These knowledge assets are related to organization members with the appropriate talent to maximize its use and reuse.
- *Talent network*. Represent the talent that exist in the organization, the organization members who possess determined talent, and the relations that exist between talent and people through the knowledge assets.
- *Roles*. Define who will be responsible for creating or updating each of the knowledge assets, who will be responsible for relating the talented people to the knowledge asset, and who can access to a specific knowledge asset.
- *Management processes*. Define how to create, update, store, access, transfer, and apply each of the knowledge assets; how to identify and relate the people with the appropriate talent to use or improve each of the knowledge assets; and how to involve the right people in software development projects.
- *Tools*. These are the software systems that will support each of the management processes, e.g. knowledge repositories, blogs, or wikis. The functionality and characteristics of these tools must be defined accordingly with the management processes that they will support.
- *Performance indicators*. Allow evaluating how the knowledge and talent management is carried out and if the goals are being achieved. These allow the organization to make the decisions about how to evolve their KT-MS.
- *Measurement processes*. Define how each of the performance indicators will be measured. The measurement processes must be designed considering the characteristics of the performance indicators.

These artifacts are interrelated; it is not possible to develop one of them without considering the characteristics of the one previously developed.

2.4. Phases and Processes.

The framework proposed by authors is divided in six phases that must be carried out by software engineering organizations in order to develop and deploy a *KT-MS*.

1. **Initial preparation.** In this phase an identification of the existing knowledge management practices is performed by a knowledge management expert in order to identify what are the existing practices and these are carried out. Based on the results of this, a training process regarding knowledge management must be

conducted to sensitize the organization's members about the importance of knowledge management.

2. **Definition of goals, scope and approach.** In this phase the organization must define its knowledge management goals and align them with its strategic goals, and also the scope and approach of it knowledge management strategy.

3. **Definition of knowledge assets and talent network.** The organization must identify all of its knowledge assets related with the areas, projects or products comprised within the scope above defined, and must identify and document who are the people possessing the expertise regarding those knowledge assets.

4. **Design of roles, responsibilities and tools.** In this phase it must be defined which members of the organization will play the following roles related with the knowledge assets as well as their responsibilities: Knowledge worker, knowledge keeper and subject matter expert (see Table 1). Also the tools required to perform such responsibilities must be designed.

5. **Design of the evaluation system.** The evaluation system is comprised of the performance indicators and the measurement process that are used to make the decision regarding how the whole knowledge management strategy must be improved along the time.

6. **Deployment of the knowledge and talent management strategy.** In this phase the strategy previously designed is deployed in the organization and the data regarding the evaluation system is collected and analyzed.

These phases obey to a general goal and have a behavior that characterizes them, as can be seen in Fig. **1**. The first phase, *Initial preparation*, will be carried out once as its goal is to prepare the software engineering organization to start the development process of the *KT-MS*.

Phases 2, 3, 4, and 5 have as general goal the development and evolution of the *KT-MS*; they are cyclical because after performing the Phase 5, the cycle will resume with the execution of the Phase 2, and thus the strategy is developed and evolved continually. Finally, Phase 6 has as its general goal to put into operation the strategy developed, or to apply the changes it can suffer over the time.

Once a software organization uses this framework to manage its knowledge and talent, the interaction between the framework and the ongoing operations and software development projects is cyclical: 1) The knowledge management strategy is developed or evolved, 2) The strategy is deployed, 3) The knowledge and talent are managed following the strategy along with the regular processes of the organization, 4) A feedback process towards the framework is carried out. Depending on the performance indicators and the evolution of the organization, the cycle could be resumed.

3. Case Study

In order to validate the usefulness of the proposed framework in the software industry, authors present a case study where it was used in a pilot project to deploy a knowledge and talent management strategy to improve the achievement of one business goal in a software engineering organization, a kind of organization where knowledge is perhaps the most valuable asset. The pilot experiment involved seventy-six Junior Software Engineers (Junior-SE) and four Senior Software Engineers (Senior-SE), eighty people in total. Senior-SEs were involved on the design and deployment of the knowledge and talent management strategy.

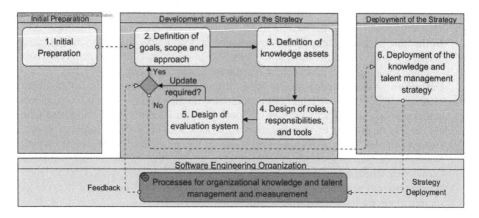

Fig. 1. Phases of the knowledge and talent management framework

Once the strategy was defined and deployed, the Junior-SEs played the roles of knowledge workers and knowledge keepers whiting the strategy. Senior-SEs played the role of subject matter experts. The case study involved the execution of the analysis and design phases of a software project; where the goal was to improve work-teams productivity through the knowledge and talent management strategy. An explanation of how each phase of the proposed framework follows.

3.1. Execution of Phase 1: Initial Preparation

This phase revealed that the organization had some practices that were carried out without the acknowledgement that they were knowledge management practices, some of these were formal such the use of documentation standards and formal modeling languages (like the Unified Modeling Language -UML- [27]), also other were less formal like coffee-meetings where minutes were taken.

As part of the execution of this phase, a ten hours training about knowledge management and the use of the proposed framework was conducted.

3.2. Execution of Phase 2: Definition of Goals, Scope, and Approach

In this phase the goal of the knowledge and talent management strategy was defined as *improve the reuse of previous experiences in the development of new projects.* This goal was related with the selected strategy goal of the organization: *Improve work-teams productivity.*

Following the proposed framework, the scope defined for the strategy was comprises of the analysis and design phase of the software development lifecycle.

And finally the approach followed was to manage the knowledge through codification, it implies make explicit the knowledge of people in order to share it across the organization.

3.3. Execution of Phase 3: Definition of Knowledge Assets and Talent Network

In this phase all the knowledge assets identified were related with the analysis and design phases of the software development lifecycle. Those assets where: the software requirement specification, use case diagrams, data base diagrams, and class diagrams. For the purpose of this case study only use case diagrams were fully analyzed.

Table 1. Knowledge management roles.

Role: Knowledge Worker	
Description	Peter Drucker introduced this term in 1959 as *"one who works primarily with information or one who develops and uses knowledge in the workplace"*[26]. Having this definition as starting point, authors considered that nowadays a knowledge worker is any member of an organization, since everyone is a potential user or creator of knowledge; every person into an organization will be considered a knowledge worker. This role could be played in conjunction with any other of the roles described in this table.
Responsibilities	• Create knowledge assets of any potential useful knowledge that they own about the services, processes, or products of the organization. • Propose improvements to those knowledge assets that they use and are subject of betterment.
Role: Knowledge Keeper	
Description	Is a person whose responsibility is to ensure the reliability of knowledge assets.
Responsibilities	• Preserve the organizational knowledge status in close collaboration with Subject Matter Experts. • Ensure that every new knowledge asset is understandable and well formed. • Purge from the organizational knowledge repository the knowledge assets that were marked to be deleted.
Role: Subject Matter Expert	
Description	This role must be played for those people in the organization who has the highest expertise in the determined knowledge assets of the organization. The person who plays this role will have the final word about the quality or usefulness of a knowledge asset if it is questioned.
Responsibilities	• Preserve the organizational knowledge assets in close collaboration with Knowledge Keepers. • Validate the suitability of every updating request for the knowledge assets of their matter of expertise.

In order to define the talent, network the people holding the highest expertise on each knowledge asset were identified as well as their direct professional relationships among the member of the organization.

3.4. Execution of Phase 4: Design roles, responsibilities, and tools.

In this phase those people holding the highest expertise on each knowledge asset were defined as subject matter experts, together with the Senior-SEs. It was decided that the role of knowledge keeper would be played at least by two people in every work-team. All people were defined as knowledge workers since it was understood that everybody had the responsibility to share and reuse knowledge. The tool defined to storage and manage the knowledge assets was a Wiki platform.

3.5. Execution of Phase 5: Design of the Evaluation System

In this phase, the evaluation system was defined to control the creation of use case diagrams. The performance of this knowledge asset was controlled with the following indicators:

- *Prod. Q.* This is the quality assessed in terms of the overall rating gave by the Subject Matter Experts of the knowledge assets (use case diagrams) developed by Junior-SEs
- *Dev. Time.* This is the development time, and indicates how many days spent the software engineers developing the knowledge asset analyzed in the pilot project.
- *Mentoring.* This is the number of mentoring sessions that Junior-SEs had with Subject Matter Experts during the development of the software product analyzed in the pilot project.

3.6. Execution of Phase 6: Deployment of the Knowledge and Talent Management Strategy

In this phase the goal of the knowledge and talent management strategy was explained to the members of the organization, and then each one was trained on the activities to be performed according to their roles. Finally, the Wiki platform was deployed and the strategy was used on three iterations of a software project.

Results obtained after the deployment of the *knowledge and talent management strategy* were encouraging when the results were analyzed. The behavior of indicators, shown that changes introduced by the use of the proposed framework were encouraging the improvement of work-teams productivity (Fig. 2).

As a preliminary statistical analysis technique, measures of central tendency were used to have a reference of the benefits of using the strategy during the pilot project execution.

As can be seen in Fig. 2Error! Reference source not found., *product quality* was drastically increased, especially if results between *Iteration 1* and *Iteration 3* are compared, this fact is more encouraging than the evolution of product quality from *Iteration 1* to *Iteration 2*, this fact highlights the potential benefits of using the designed strategy as a guide to manage organizational knowledge and talent to achieve business goals measured in terms of indicators aligned with them. Additionally, *development time* decreased without compromising product quality, this is a very promising fact, especially for organizations, like in software industry where any day saved during a project implies a potential benefit like the improvement of client satisfaction. The *number of mentoring sessions* also decreases, meaning a better use of the existing explicit knowledge on the Wiki platform. Following a deeper statistical analysis of control variables is presented; it was performed to corroborate if there were any correlations among control variables to verify if any change on their values could really be significant for their behavior and evolution.

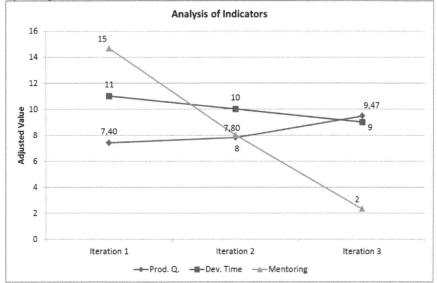

Fig. 2. Evolution of indicators on Iterations 1, 2 and 3.

3.7. Statistical Validation of Case Study Indicators

Since achieving business goals through a better management of organizational knowledge and talent is the main objective of authors' proposal, authors wanted to know if the positive effect shown in Fig. **2**, where product quality was not compromised when mentoring sessions and development time decreased, was a matter of fortuitousness or there is any relationship among the variables. Authors conducted a statistical analysis of the behavior of control variables during the three iterations in order to find out if there was any interdependence among the variables with statistical significance.

Since the distribution of the case study results was not normal, in order to verify if there were relations among the control variables a correlation analysis was conducted using the Spearman's Rho test [28] using the SPSS software [29] and the data from the three iterations. Following Tables 2 to 4 show the results of the Spearman's Rho test.

As it can be observed from the analysis of Spearman's Rho test, there is not a significant statistical relationship among any of the indicators, there is a relation among them nevertheless is too weak; however from an analytic point of view, there is important to notice the facts show before, according to the proposed strategy the organization improved their capability to manage their knowledge and talent, and also, there are indications that suggest that having more knowledge assets represented as explicit knowledge, help the organization to improve their management of time in product development and experts in mentoring sessions without compromising the quality of the developed product. Nevertheless, there are some questions in the air, what hidden organizational aspects are promoting apparent improvement on knowledge management capabilities of the organization? Is there a need of a different mechanism to assess the goals achievement beyond the indicators? Or these statistical results are a suggestion that the benefits from knowledge must be analyzed analytically instead of statistically? Authors are aware that further research is needed to clear these questions, and they are part of the future works derived from this exploratory work to align knowledge management initiatives with business goals achievement.

Table 2. Spearman's Rho test results for Iteration 1.

	ProdQ1	DevTime1	Mentoring1
ProdQ1	-	r_s=0.080 p=0.492	r_s=-0.115 p=0.320
DevTime1	r_s=0.080 p=0.492	-	r_s=-0.003 p=0.980
Mentoring1	r_s=-0.115 p=0.320	r_s=-0.003 p=0.980	-

Table 3. Spearman's Rho test results for Iteration 2.

	ProdQ2	DevTime2	Mentoring2
ProdQ2	-	r_s=0.221 p=0.055	r_s=-0.211 p=0.067
DevTime2	r_s=0.221 p=0.055	-	r_s=0.024 p=0.838
Mentoring2	r_s=-0.211 p=0.067	r_s=0.024 p=0.838	-

Table 4. Spearman's Rho test results for Iteration 3.

	ProdQ3	DevTime3	Mentoring3
ProdQ3	-	r_s=0.060 p=0.605	r_s=0.119 p=0.305
DevTime3	r_s=0.060 p=0.606	-	r_s=0.048 p=0.682
Mentoring3	r_s=0.119 p=0.305	r_s=0.048 p=0.682	-

4. Expected Benefits and Future Work

The expected benefits of this proposal are for both, industry and academia. The industry can use the proposed framework in order to understand how mature is in managing its organizational knowledge and talent, identify why its previous knowledge management initiatives did not bring the expected results, and achieve an effective management of its organizational knowledge. For the academia, given the breadth of the knowledge and talent management field, and that so much research is still required in aspects, such as knowledge elicitation or knowledge transfer in global software development processes; it is expected that this framework serves as a reference, showing the whole picture of managing knowledge and talent in a software engineering organization, and allowing to understand where and why a specific aspect being investigated is located within the whole picture, and how it is related to other aspects.

Authors' future work focuses on refining the structure of the artifacts of a *KT-MS*, and the resources used to assist in the development of those artifacts; all in collaboration with various software companies interested in being able to manage their organizational knowledge and talent.

Cultural and emotional factors affecting knowledge and talent management are very important for the effective deployment of the proposed framework. In many cases it is assumed that factors like leadership, organizational culture, and organizational trust have been previously addressed by organizations, however these factors are not always considered in organizations, this is mandatory and must be considered as part of the framework deployment success and technologically supported if possible, so that further research to improve this proposal would take into account these factors.

References

[1]   I. Rus and M. Lindvall, "Knowledge management in software engineering," *IEEE Softw.*, vol. 19, no. 3, pp. 26–38, May 2002.

[2]   A. Tiwana, "An empirical study of the effect of knowledge integration on software development performance," *Inf. Softw. Technol.*, vol. 46, no. 13, pp. 899–906, Oct. 2004.

[3]   P. Fehér and A. Gábor, "The role of knowledge management supporters in software development companies," *Softw. Process Improv. Pract.*, vol. 11, pp. 251–260, 2006.

[4]   R. Patnayakuni, A. Rai, and A. Tiwana, "Systems development process improvement: A Knowledge integration perspective," *IEEE Trans. Eng. Manag.*, vol. 54, no. 2, pp. 286–300, May 2007.

[5]   F. O. Bjørnson and T. Dingsøyr, "Knowledge management in software engineering: A systematic review of studied concepts, findings and research methods used," *Inf. Softw. Technol.*, vol. 50, no. 11, pp. 1055–1068, Oct. 2008.

[6]   T. H. Davenport and L. Prusak, *Working knowledge: how organizations manage what they know*. Harvard Business Press, 2000, p. 240.

[7]   M. Buckingham and C. Coffman, *First, break all the rules: What the world's greatest managers do differently*. Simon & Schuster, 1999, p. 255.

[8]   M. Alavi and D. E. Leidner, "Review: Knowledge management and knowledge management systems: conceptual foundations and research Issues," *MIS Q.*, vol. 25, no. 1, pp. 107–136, 2001.

[9]   K. D. Joshi, S. Sarker, and S. Sarker, "Knowledge transfer within information systems development teams: Examining the role of knowledge source attributes," *Decis. Support Syst.*, vol. 43, no. 2, pp. 322–335, Mar. 2007.

[10] A. Aurum, F. Daneshgar, and J. Ward, "Investigating knowledge management practices in software development organisations – An Australian experience," *Inf. Softw. Technol.*, vol. 50, no. 6, pp. 511–533, May 2007.

[11] L. Mathiassen and P. Pourkomeylian, "Managing knowledge in a software organization," *J. Knowl. Manag.*, vol. 7, no. 2, pp. 63–80, 2003.

[12] H. Lee and B. Choi, "Knowledge management enablers, processes, and organizational performance: An integrative view and empirical examination," *J. Manag. Inf. Syst.*, vol. 20, no. 1, pp. 179–228, 2003.

[13] G.-W. Bock, R. Sabherwal, and Z. Qian, "The effect of social context on the success of knowledge repository systems," *IEEE Trans. Eng. Manag.*, vol. 55, no. 4, pp. 536–551, Nov. 2008.

[14] R. McDermott, "Why information technology inspired but cannot deliver knowledge management," *Calif. Manage. Rev.*, vol. 41, no. 4, pp. 103–117, 1999.

[15] U. Kulkarni, S. Ravindran, and R. Freeze, "A knowledge management success model: Theoretical development and empirical validation," *J. Manag. Inf. Syst.*, vol. 23, no. 3, pp. 309–347, Jan. 2007.

[16] T.-C. Chang and S.-H. Chuang, "Performance implications of knowledge management processes: Examining the roles of infrastructure capability and business strategy," *Expert Syst. Appl.*, vol. 38, no. 5, pp. 6170–6178, May 2011.

[17] B. D. Janz and P. Prasarnphanich, "Freedom to cooperate: Gaining clarity Into knowledge integration in information systems development teams," *IEEE Trans. Eng. Manag.*, vol. 56, no. 4, pp. 621–635, Nov. 2009.

[18] C. Ebert and J. De Man, "Effectively utilizing project, product and process knowledge," *Inf. Softw. Technol.*, vol. 50, no. 6, pp. 579–594, May 2008.

[19] E. Coakes, A. D. Amar, and M. L. Granados, "Knowledge managemer for the twenty-first century: A large comprehensive global survey emphasizes KM strategy," in *European and Mediterranean Conference on Information Systems*, 2009, vol. 2009, pp. 1–13.

[20] N. Mehta, "Successful knowledge management implementation in global software companies," *J. Knowl. Manag.*, vol. 12, no. 2, pp. 42–56, 2008.

[21] M. H. Zack, "Developing a knowledge strategy," *Calif. Manage. Rev.*, vol. 41, no. 3, pp. 125–145, 1999.

[22] M. E. Greiner, T. Böhmann, and H. Krcmar, "A strategy for knowledge management," *J. Knowl. Manag.*, vol. 11, no. 6, pp. 3–15, 2007.

[23] A. Saito, K. Umemoto, and M. Ikeda, "A strategy-based ontology of knowledge management technologies," *J. Knowl. Manag.*, vol. 11, no. 1, pp. 97–114, 2007.

[24] S. Iuliana, "A knowledge management practice investigation in Romanian software development organizations," *WSEAS Trans. Comput.*, vol. 8, no. 3, pp. 459–468, 2009.

[25] C. Barclay and K.-M. Osei-Bryson, "An exploration of knowledge management practices in IT projects : A case study approach," *Proc. Sixt. Am. Conf. Inf. Syst. Lima, Peru*, 2010.

[26] P. F. Drucker, *Managing in a Time of Great Change*. Plume, 1998, p. 384.

[27] R. Miles and K. Hamilton, *Learning UML 2.0*. O'Reilly Media, 2006, p. 288.

[28] J. S. Maritz, *Distribution-Free Statistical Methods*, 2nd ed. London, UK: Chapman and Hall, 1995, p. 256.

[29] IBM, "IBM SPSS software," 2012. [Online]. Available: http://www-01.ibm.com/software/analytics/spss/. [Accessed: 25-Mar-2016].

A method for automatic generation of explanations from a Rule-based Expert System and Ontology

Victor Flores[1], Yahima Hadfeg[1], Juan Bekios[1], Aldo Quelopana[1], Claudio Meneses[1]

[1] Department of Computing & Systems Engineering, Universidad Católica del Norte, Angamos Av, 0610,Antofagasta, Chile
{vflores , Yahima.Hadfeg01, jbekios, aldo.quelopana, cmeneses}@ucn.cl

Abstract. Expert systems (ES) usually generate extensive inference-trees before showing to users a definitive result related to a complex dynamic system (DS) behavior. These inference-trees are not included in the results but it could provide additional information to understand the overall performance of a DS. They contain a set of statements that describe the knowledge about the truths of the DS plus a set of constrains that can give statements that must be true in the DS behavior. This document describes a method to generate explanations based on the conclusions reached by an ES respect to the DS behavior, using a specific ontology and discourse patters. The input of the method is an intermediate-state tree (the inference-tree) and a specific knowledge-domain represented by the ontology. The document describes the software architecture to generate the explanations and the testing cases designed to validate the results in a complex real domain, such as the copper bioleaching domain.

Keywords: Expert System, data-to-text system, intelligent multimedia presentation systems, interactive data analysis, knowledge representation for dynamic systems.

1 Introduction

Complex Expert Systems usually generate large inference trees as intermediate-output before present results to users; these intermediate-output can be intermediate states and they can helps to improve the comprehension of Expert System results; but normally, this task are not made because it can be complicate and difficult to show. The management of dynamic systems usually involves operator teams and decision makers , and in general the human teams are interested not only in the output of an expert system but also in the description of possible states of the dynamic system in the time. An example of a dynamic system is an industrial copper bioleaching heap; an industrial bioleaching heap is divides into lifts and strips. Sometimes the heap are building with run-of-mine (ROM) ore characterized as low-grade sulfide material . Bioleaching is the process that uses ROM and microorganisms' populations to obtain valuable metals (e.g. copper) with otherwise that would not be economically profitable , [4]. Various works are presented in the literature that shows results of bacteria incidence on optimum conditions of bioleaching.

© Springer International Publishing AG 2017   167
J. Mejia et al. (eds.), *Trends and Applications in Software Engineering*, Advances
in Intelligent Systems and Computing 537, DOI 10.1007/978-3-319-48523-2_16

Many mathematical models are making to support the bacterium behavior on simulated bioleaching heap, but on real heap there are not adequate duo to the complexity and limited knowledge respect to bacterial population behavior. In this context some expert systems can help operators to understand the copper bioleaching process and support the making decision process . A tool that automatically generates descriptions of a heap behavior can help end-user to understand and analyze the heap complexity. Copper mining is the most important economic activity in Chile, generating almost a 10% of worldwide copper production .

This document describes an innovative method to generate explanations from the conclusions arrived by an expert system on bioleaching copper process. Our method is able to generate text and graphics descriptions in order to generate a multimedia explanation. The method use specific domain ontology and discourses patters to generate the explanation. We validated our method and showed partial results in the domain of copper bioleaching that work with an expert system developed by the Centro de Biotecnología "Profesor Alberto Ruiz" (CBAR)[1], Chile.

In the following, we describe the method and the software methodology. First, we describe the software faces applied to the software building, next we describe the knowledge representation and other representation of the discourse structures, the inference abstractions method and the presentation planning. We also describe how we validate our method and, at the end of this paper we make a discussion about the practical utility and a comparative discussion about the human interpretation of expert system results, and the automatically generated presentation by the software.

2. Software methodology and work context

Software has emerged as a means for creating value to products and services in many industries include the mining copper industry. In this context, it is well known that quality software products depend largely on the processes development, final quality and level of knowledge used to build the responses . The software building was guided by the steps described on Fig. 1. The methodology corresponds to an incremental process that combines software engineering with artificial intelligence steps in order to generate the prototype.

Fig. 1 details eight steps, **the first** one was made society of experts of CBAR to identify how to show relevant information, relevant concepts, among other related to mining copper domain. The **second step**, structure of inference-tree from the bioleaching expert system (BES) was identified. This inference structure is intermediate result of the BES; a framework consist of Java Eclipse[2] and Drools[3] was been used to build the BES, and experts of CBAR were consulted to know how data were important.

[1]	Biotechnology center, http://www.ucn.cl/sobre-ucn/vicerrectorias-ucn/vicerrectoria-de-investigacion-y-desarrollo-tecnologico/centro-de-biotecnologia/
[2]	http://www.eclipse.org/home/
[3]	http://www.drools.org/

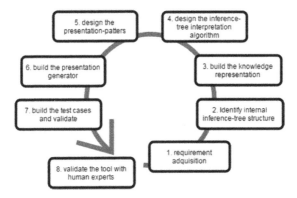

Fig. 1. Methodology designed by build the explanations generator.

The BES input are both data from sensors and information about heap status. Before generate the output the BES produce an inference-tree, the structure of each node is *<value,object,description,recommendation>* where *value* is a variable value (e.g. temperature), *object* is an object of the system (e.g. heap, bacteria, etc.), *description* is a partial conclusion that BES arrived during the inference process and *recommendation* is a suggestion related to *description;* both *description* and *recommendation* can be void-value in a particular node.

The **third step** was to develop the specific ontology with the collaboration of CBAR experts. The core of this proposal is made during **fourth** to **sixth** steps, being the steps **seven** and **eight** to validate it.

A formal definition of an ontology proposed by Gruber [17] is that "an ontology is an explicit specification of a conceptualization", an ontology model the domain using the elements: concepts, attributes and relationships in order to specify the domain vocabulary. In the literature various ontologies classifications are presented, Mizoughi and colleagues [18] propose that ontology can be a domain, task or general/common specification. Ontology representation is comprised of four main elements [21]: *concept* or *class* as an abstract group, set or collection of elements; *instance* is the "ground-level" component, *relation* or *slot* is used to express relationships between two concepts and *axiom*, used to improve constraints on the values of classes or instances. These elements are combined on tuples in order to generate the structure of an ontology [22].

Regarding the work context, metals are one of the most significant natural resources for the humans. The metals are extracted from the rocks using different techniques, and several produce environmental pollution [7] [8]. New and cleaning forms of copper extraction are available such as bioleaching that gives the option of obtaining metals such as copper from low-grade material, at a lower cost and minimized environmental impact [3].

In order to support many activities on copper bioleaching industry (i.e. decision making), many copper mines are using expert systems. An expert system can be described by a computer system that operates by applying an inference mechanism to obtain results, based on a specific knowledge represented and that results are similar to obtained for an expert human [20]. The work described in [20] presents a rules-based expert system for analyze and predicting consumer preference of a new product using decision making processes.

For an Expert System, the knowledge representation can be made with several Artificial Intelligence techniques [8], one of these are ontologies. Ontologies provide high level of representativeness allowing wealth description structures on particular domains [7]. Actually, ontologies are using to represent and store information, and also to search and retrieve information. The ontological representation and Expert System combination is frequently used in order to provide software solutions, for example, the work in [19] describes a rule-based algorithms and an ontology-based context model to provides personalized services of E-Learning for disabled students in higher education.

Deterministic rules are commonly used in Expert Systems as a valid form to knowledge representation [8] and to infer Dynamic System states , [9]. It is due to many real complex situations are governed by deterministic rules and this representation can simplify the problem complexity and bring with human reasoning. These reasons have contributed to rules popularity on different domains. The kind of Expert System type considered on this proposal is the Rule-based Expert System (RBES); it's can have different states over time and those states are measurable with a variety of devices such as electronics devices, specific sensors, etc. There are previous works to summarize behavior or relevant information, for example in a model for automatically generation of presentations is presented.

Today there are several techniques to knowledge representation. A classification of knowledge base modeling techniques is presented on [11], this classification is made based on the fundamental theories of knowledge based modelling and manipulation. On this classification, the following groups are considered: linguistic knowledge based [12], expert knowledge bases [13], [14], ontology and cognitive knowledge bases [15], [16]. Actually, these knowledge representations are widely used for support Intelligent Systems build applications.

3. Proposal

This method is proposed as a tool for generating explanations about an expert system result. The input to our method is both the expert system result and the inference tree generated by the expert system inference-tree (Fig. 2 right).

The presentation plan contains the directives (text and graphs patterns) to build the presentation. Fig. 2 describes the method architecture; there are three principal processes, for the inference-tree interpreter the inputs are both the inference-tree where each node has the structure describe as: <*var, value, object, description, recommendation*>, and the specific ontology. This ontology has the relevant knowledge to decide how interpretations could be select for the relevant behavior generator process, for later it can be provided the explanations generator process. The explanations generator use the ontology and the presentation model in order to produce the presentation plan and the output are explanations.

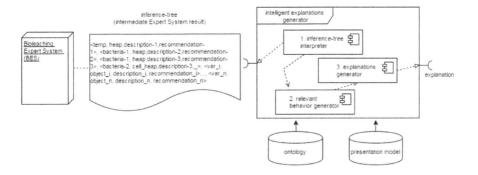

Fig.2. Mean components of the method architecture.

In order to represent the domain knowledge an ontology schema was been designed. There are two aims of the ontology construction, the first being to improve an usable representation closed to important dynamic system element's to be described and the second to ensure that its concepts can be understood by the presentation module. The ontology is composed by structural and behavior elements and this is explained in the section called Case study.

3.1. The presentation model

The presentation model contains the criteria to build automatically explanations about the expert system results; these descriptions have the dynamic system characteristics that can helps to understand the expert system results. To do it, the presentation model combines forms of presentation such as text in natural language and graphic elements for illustrate temporal series values.

The goal of this presentation model is to improve the visibility of the expert system results, adding abstract information from the inference-tree. Multimedia presentations are a suitable schema for this purposes and it provides an efficient way to display complex information from a data source [1]. The way designed to outline the explanations is through a presentation plan that specifies how to present information. The presentation plan is designed as a combination of *elements of presentation* and *presentation strategies*.

The elements of presentation are explanatory elements that we used to show the expert system status and explain salient events; specifically in this proposal presentation elements are text-template and graphic-directive. A text-template combines variables (concept characteristic) and fixed text, and the graphic-directive are commands to build graphics within the explanation.

Presentation strategies are patterns that are schematics of how to articulate the presentation; a presentation strategy establishes relationships between text-template and graphic-directive and indicates which of these are to be used to structure the presentation.

Presentation strategies are selected using an inference process based on rules. In order to represent the rules predicate logic was used and a pseudo-representation based on prolog was made, these rules are saved in the presentation model knowledge-base, the following are examples of these rules: mesophile *is-a* Bacteria, ther-mophilic *is-a* Bacteria, temp is the temperature value and oxid-level is the oxidation level value.

IF value(mesophile)>10e5, value(thermophilic)>10e6, value(oxid-level)<5 THEN select-strategy(strategy-1).

3.2. Case study

To validate the model we used a real world domain with enough complexity, avail-able data and an Expert System that generate interpretations of the system behavior. A bioleaching heap is a real dynamic system that have certain features as follows: (1) a large and complex amount of physical elements that might have also a significant number of characteristics and connections between them, (2) the real dynamic system have to be managed and supervised by human experts, and (3) these experts can inter-act with the system in order to change its behavior.

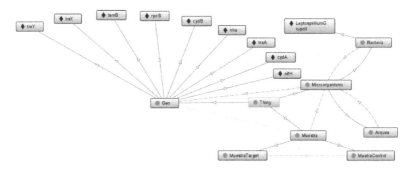

Fig. 3. Ontology using Protégé.

The ontology has tree parts for three modules of the expert system. Fig. 3 describes the gene expression ontology; it consists of 6 main classes and this was created using the Protégé tool [10], a brief description of these elements and their characteristics is: Gen represent a biological gene, Muestra represent a biological sample, Bacteria rep-resent a bacterium, Arquea represent a archaea, Microorganismo represent a microor-ganism, MuestraTarget represent a target sample and MuestraControl represent a samplet that is used to contrast with the MuestraTarget sample. The relations between classes were been defined as following: Gen *is-a* Thing, Microorganismo *is-a* Thing, Bacteria *is-a* Thing, Muestra *contains* Gen, MuestraTarget *contains* Gen, MuestraControl *contains* Gen, Bacteria *is-a-kind-of* Microorganismo, Arquea *is-a-kind-of* Microorganismo, MuestraControl *is-a-kind-of* Muestra, MuestraTarget *is-a-kind-of* Muestra. Axioms on this representation are: **Axiom-1:** if B *is-a* Bacteria and A *is-a* Arquea → A≠B and A∩B= Ø, **Axiom-2:** if A *is-a* MuestraTarget, B *is-a* MuestraControl → A≠B and A∩B= Ø, **Axiom -3:** if B *is-a* Bacteria and A *is-a* Ar-quea → A⊆Microorganismo and B⊆Microorganismo and A∪B= Microorganismo,

Axiom-4: if C, D *is-a-kind-of* Muestra, C *is-a* MuestraControl, D *is-a* MuestraTarget
→C≠D and C£D, where £ represent an expression of D respect to C and £ value
IN {over-expressed, sub-expressed, un-expressed, equal-expressed}.

Ontology contains also abstract qualitative values, in order to abstract qualitative
values specific functions were been defined, some examples of this functions are: *is-
a-abstract-of*, this mean when two or more instances of a concept are abstracting by
other more general. For example, if A is the set of elements {A, B, C} and there are related with other element D by *is-a relation* relation, then D represent all the A elements. For run this relation an algorithm was been defined as follow:

```
Input: set of elements A.

Output: single element that abstract A. Define: n (in-
teger), abstrae (Boolean) as local variables

Procedure:(1) Select all similar elements and set in A;
(2) For each Ai ∈ A = [A1, A2, Ai.., An] where n≥2; (3)
If ((Ai is-a D) && (D is-a-abstract-of Ai)) them ab-
strae ← TRUE    Else abstrae ← FALSE; (3.1) IF (abstrae)
then Return (D)   Else Return (A)
```

The behavior elements defined on the ontology are state, these states can be related
to a concept or a set A of concepts, and in this case stare is a representative state of A.
An example of this kind of relation is: *is-a-abstract-state-of*, when the behavior of
two or more instances is abstracting by one more general behavior description related
to other instance. For example, if $A=\{$(E1, S1), (E2, S2), .., (Ei, Si), (En, Sn)\}$ is the
set of elements where each one (Ei) have a state related (Si), and $1 \leq i \leq n$, then the behavior of these elements can be abstain in a representative element with a representative state, using the *is-a-abstract-state-of*. For run this relation an algorithm was been
defined as follow:

```
Input: set A(Ei, Si) where i≥2, A'=(null, null). Out-
put: set A'(Ej, Sj) where j=1. Define: n,i (integer),
abstrae-state (Boolean) as local variables

Procedure: (1) Select all similar Ei  in set A  where
each Ei have the Si state, (2) For each (Ei, Si) ∈ A,
(3) If ((Ej is-a-abstract-of Ei)&& (Sj Is-a-ab-
stract-state-of Si) them abstrae-state ← TRUE    Else
abstrae ← FALSE , (3.1) IF (abstrae) then      Return
(A') Else Return (A)
```

In add, to generate the explanation a specific presentation model was created. The
events: internal/external heap temperature, copper oxidation level, gene expression in
the control/target sample, no-response available from the expert system and no-ex-
planation were considered; twenty different text-elements, seven graphic-elements
and seven presentation strategies were been constructed. Table 1 contains examples of
element of presentation and table 2 show variables values that trigger presentation
strategies. Fig. 4 shown a prototype of explanation that combine text and graphic.

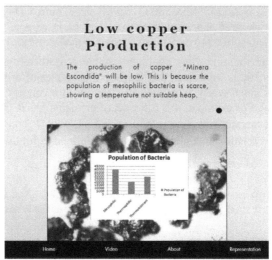

Fig 4. Web page prototype to display the explanation.

Table 1. Elements of presentation for the bioleaching domain.

Element of presentation	Description
oxidation_header(X)	Text structure to show a headline, this text abstracts the oxidation state <X>, where X is an oxidation appreciation value. X in {high, medium, low}.
text_oxidation_detail1(X, T)	Text structure that show a detail related to oxidation level and details of variables related to this situation, where X is the oxidation value, T is the temperature value and X,T in N
text_oxidation_detail2(X, T)	Text structure to show a detail related to oxidation level and details of variables related to this situation, where X is the oxidation value. T is the temperature value and X, T are in N. This structure added a copper recuperation interpretation based on the previous values.

Table 2. Presentation strategies for the bioleaching domain.

Bacteria groups	Temperature	Oxidation level	Presentation strategy
(mesop>10e5)∧ (therm <10e4)	heap_temp>=24	heap_oxi>10	Strategy_1: oxidation_header("High oxidation level on the heap"), text_oxidation_detail1("the oxidation level is 10, the heap temperature is greater than 24 °C")
(mesop>10e5)∧ (therm <10e4)	heap_temp<24	heap_oxi<=10	Strategy_1: show_header("Low oxidation level on the heap"), text_detail("the oxidation level is 10, the heap temperature is less than 24 °C)

4. Conclusions

This work is a step towards creating a module that automatically generates explanations using a domain ontology and knowledge base specific to the task of bioleaching of copper in Chile. This work differs from previous work (1) in that it incorporates an ontological representation that provides more richness and flexibility and (2) use process data analysis of the domain processes to identify the more relevant behaviors to show in the explanation. An ontological model provides flexibility to representation, unlike previous work using first-order logic or other methods with less flexibility. This paper also describes the use of discourse patterns as an effective way of build text explanations from data or tree-inference values.

The work done so far has allowed validate that ontologies are a technique for representing knowledge suitable to the specific domain, which can be used easily for the purposes of this proposal. This in contrast to other techniques of knowledge representation, such as a semantic network, which do not have the representative capability for the kind of situations related to the domain of copper bioleaching.

This proposal is an improvement of previous similar works. Unlike other proposals, this algorithm generates explanations of intermediate states of an expert system. Under this context, the expert system can generate inaccurate interpretations related to the state of the dynamic systems, but this information can be complemented by explanations of the inference and intermediate states as those generated using this method. These mentioned explanations are more closed to human experts because they are provided using a natural way, and the information quality is incremented. Thus the practical utility of the expert system is improved, and users can understand better the corresponding result.

References

1. Molina M., Flores V.: Generating multimedia presentations that summarize the behavior of dynamic systems using a model-based approach. Expert Syst. Appl. 39(3), pp. 2759–2770 (2012)
2. Demergasso D. Galleguillos F., Soto P., Seron M., Iturriaga V.: Microbial succession during a heap bioleaching cycle of low grade copper sulfides: Does this knowledge mean a real input for industrial process design and control. Hydrometallurgy. 104(3), 382–390 (2010)
3. Soto P,. Galleguillos P., Seron M.,. Zepeda V, Demergasso C., Pinilla C.: Parameters influencing the microbial oxidation activity in the industrial bioleaching heap at Escondida mine, Chile. Hydrometallurgy, 133, 51–57 (2013)
4. Kaibin F., Hai L., Deqiang L., Wufei J., Ping Z.: Comparsion of bioleaching of copper sulphides by Acidithiobacillus ferrooxidans. African J. Biotechnol. 13(5), 664–672 (2014)
5. Data P.: ICSG PRESS RELEASE Date Issued : 20th December 2013 Copper : Preliminary Data for September 2013, 00(September 2013) (2013)
6. Mejia J., Muñoz E., Muñoz M.: Reinforcing the applicability of multi-model environments for software process improvement using knowledge management. Sci. Comput. Program., 121, 3–15 (2016)
7. Abdel-Fattah T. M., Haggag S. M. S., Mahmoud M. E.: Heavy metal ions extraction from aqueous media using nonporous silica. Chemical engineering journal, 175, 117-123 (2011)

8. Khaliq A., Rhamdhani M. A., Brooks G., Masood S.: Metal extraction processes for electronic waste and existing industrial routes: a review and Australian perspective. Resources, 3, 152-179 (2014)
9. Green N., Carenini G., Kerpedjiev S., Mattis J., Moore J., Roth S.: AutoBrief: an Experimental System for the Automatic Generation of Briefings in Integrated Text and Information Graphics. International Journal of Human-Computer Studies, 61(1), 32–70, (2004)
10. Gennari J., Musen M., Fergerson R., Grosso W., Crubezy M., Eriksson H., Noy N., Tu S.: The evolution of Protégé: an environment for knowledge-based systems development. International Journal of Human Computer Studies, 58(1), 89–123 (2003)
11. Bimba A. T., Idris N., Al-Hunaiyyan A., Mahmud R., Abdelaziz A., Khan S., Chang V.: Towards knowledge modeling and manipulation technologies: a survey. On International Journal of Information Management, 36(6), 857-871 (2016)
12. Baker C.F.: FrameNet: a knowledge base for natural language processing. On the proceedings of frame semantics in NLP: a workshop in honor of Chuck Fillmore (2014)
13. Driankov D., Hellendoorn H., Reinfrank M.: An introduction to fuzzy control. Springer Science & Business Media (2013)
14. Kerr-Wilson J., Pedrycz W.: Design of rule-based models through information granulation. Expert Systems with Applications, 46, 274–285 (2016)
15. Sánchez D., Moreno A.: Learning non-taxonomic relationships from web documents for domain ontology construction. Data and Knowledge Engineering, 64(3), 600–623 (2008)
16. Sicilia M.-A.: Handbook of metadata, semantics and ontologies. World Scientific (2014)
17. Gruber T.R.: A translation approach to portable ontology specifications. Knowledge Acquisition, 5, 199-220 (1993)
18. Mizoguchi R., Vanwelkenhuysen J., Ikeda M.: Task ontology for reuse of problem solving knowledge. In Very Large Knowl. Bases Knowl. Build. Knowl. Shar, 46–59 (1995)
19. Ongenae F., Claeys M., Dupont T., Kerckhove W., Verhoeve P., Dhaene T., De Turck F.: A probabilistic ontology-based platform for self-learning context-aware healthcare applications. Expert Syst. Appl., 40(18), 7629–7646 (2013)
20. Yang Y., Fu C., Chen Y., Xu D., Tang S.: A belief rule based expert system for predicting consumer preference in new product development. Knowledge-Based Systems, 94, 105-113 (2016)
21. Mustafa Taye M.: Undestanding Semantic Web and Ontologies: Theory and Applications. Journal of Computing, 2(6), 182-192 (2010)
22. Neches R., Fikes R., Finin T.W., Gruber T.R., Patil R.S., Senator T.E., Swartout W.R.: Enabling Technology for Knowledge Sharing. Presented at AI Magazine, 36-56 (1991)

Decision-Support System for Integration of Transactional Systems and Analytical Models in the Pharmaceutical Industry

Edrisi Muñoz[1], Elisabet Capón-García[2], José Miguel Laínez-Aguirre[3],
Antonio Espuña[4] and Luis Puigjaner[4],

[1] Centro de Investigación en Matemáticas A.C., Jalisco S/N, Mineral y Valenciana 36240, Guanajuato, Mexico
[2] Department of Chemistry and Applied Biosciences, ETH Zurich, Wolfgang-Pauli-Str 10, 8093 Zurich, Switzerland
[3] 342 Bell Hall, Department of Industrial and Systems Engineering, University at Buffalo, Amherst, NY 14260, USA
[4] Department of Chemical Engineering, Universitat Politècnica de Catalunya ETSEIB, Avda. Diagonal, 647, E-08028 - Barcelona, Spain

elisabet.capon@chem.ethz.ch, emunoz@cimat.mx, jmlainez@gmail.com, {antonio.espuna, luis.puigjaner}@upc.edu

Abstract. Pharmaceutical industry is a highly competitive and global business, which requires sophisticate tools efficient decision-making. Decision-support tools rely on both robust information systems and accurate analytical tools capable of capturing and solving engineering and management problems. This paper presents a decision-support framework that aims to bridge the gap between transactional and analytical systems for the pharmaceutical industry. Specifically, the developed framework allows creating information quality from existing information systems, to automatically deliver it to the optimization models, and to provide optimization results for final implementation.

Keywords: Knowledge Management; Hierarchical Integration; Scheduling; Semantic Model; Information Systems.

1 Introduction

The development of pharmaceutical industry has spawned many advances, while acquiring a high level of complexity. Hence, tackling existing challenges entails considering the perspective of the company as a whole. Enterprise-wide decision problems arising in the pharmaceutical industry can be broadly classified in three categories [1], namely product development pipeline management, capacity planning, and supply chain management (SCM). The last category encompasses operational aspects of manufacturing and delivering the product to the patient. As pointed out by [2], time-to-market is certainly the most important driver in the pharmaceutical industry. Hence, the plant responsiveness becomes critical and a daily issue in pharmaceutical plants consists of optimally fulfilling customers' demands by managing production orders and accommodating them to the available resources.

© Springer International Publishing AG 2017   177
J. Mejia et al. (eds.), *Trends and Applications in Software Engineering*, Advances
in Intelligent Systems and Computing 537, DOI 10.1007/978-3-319-48523-2_17

The increased availability of enterprise information and rapid escalation in computing power have made it possible to deploy decision support tools, including models, simulations and optimization frameworks which can facilitate informed decisions that consider the interdependencies among the functional units and the necessary integration of decision levels [1]. In this sense, robust information systems (IS) are extremely functional for relying on decision-making in pharmaceutical companies. Generally, Enterprise Resource Planning (ERP) systems are implemented in practice. These systems help to provide accurate information at the right time for operational decisions, thus influencing the performance of the management and ultimately improving decision making within the firms. Even more, pharmaceutical companies have found direct impacts on the end-users' performance and productivity. The main features of an effective implementation of IS are the flexibility, traceability, accessibility, and ease of use. In addition, a clear organizational structure, a high ability for management change, and shared knowledge and technology infrastructure are extremely important for ensuring an efficient IS implementation [3].

ERP systems storing a huge amount of data related to the enterprise transactions are commonly available. However, the effective and timely use of rigorous optimization models still require robust and reliable data acquisition systems to extract useful information and to drive the proposed enterprise wide coordination strategies [4]. Therefore, the linking functionality between transactional and analytical models is still missing in the current information systems [1]. The current reality at most large pharmaceutical companies is that they are "data-rich and information poor" [5]. Indeed, the time required to assemble the required data for an enterprise model and to insure its validity very probably far exceeds the time and effort spent in creating the models. One alternative to support enterprise optimization models consists of integrating process information systems using process and business ontologies. Precisely, ontologies have been proposed in the literature for establishing linkages between data and models, specifically ontologies for pharmaceutical engineering [6,7,8], ONTOCAPE [9,10], and an ontological framework for scheduling, supply chain planning and hierarchical integration [11,12,13]. A different alternative presented by [5] extends industry standards (ANSI/ISA-88.xx & ANSI/ISA-95.xx), originally developed for process plant data, to be used in analytical and business enterprise processes. As a result, modeling the process plant, as well as the analytical and business enterprise data together creates a framework with the potential to deliver enhanced knowledge management. Nevertheless, data standards alone will not achieve this ultimate goal, and more sophisticated modeling approaches that can be plugged into the analytical side of the data warehouse are required [1].

This work presents an ontological framework that captures the pharmaceutical environment using industry standards, namely ANSI/ISA-88 and ANSI/ISA-95, and further connects to analytical models for improving decision-making. Specifically, the scheduling function of the enterprise will be tested in this work. Section 2 describes the methodology, including the standards, the ontological framework and the plant scheduling problem. Section 3 presents the application of the framework to the scheduling of a multipurpose batch plant operating in campaign mode. Finally, Section 4 summarizes the main conclusions derived from the work and outlines further developments.

2 Methodology

This work proposes the application of an ontological framework relying on ANSI-ISA standards in order to improve the efficiency of information exchange for aiding decision-support tools in the pharmaceutical environment. Section 2.1 presents the features of ANSI-ISA standards, and their role in this contribution. The backbone of the framework lies in the creation of an Operations Research (OR) semantic model capable of capturing the nature of problems and technologies for decision making in the enterprise [15]. This OR ontology (ORO) is integrated with other two semantic models which were previously developed, namely the Enterprise Ontology Project (EOP) [16] and the Ontological Math Representation (OMR) [17], thus enhancing the functionalities of the original ontological framework (Fig. 1). The scope of these models comprises the representation of the real system for EOP, the mathematical representation domain for OMR, and finally the problem design representation for ORO. Table 1 summarizes the different ontological models, their domain and their metrics, namely classes, data and object properties.

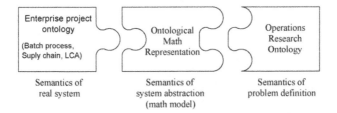

Fig. 1. Relationship among the different domain semantic models

Table 1. Summary of the ontological models and their metrics (classes and properties)

Model name	Domain	Short description	Metrics	Num
Enterprise ontology model (EOP)	Control; Automation; Monitoring; Scheduling; Planning; Supply chain; Life cycle assessment	Base line for representing physical, process, procedure, recipe and environmental models of a process plant	Classes / Object prop. / Data prop.	284 / 237 / 40
Ontological math representation (OMR)	Mathematical equations	Base line for representing logic and algebraic mathematical equations based on mathematical elements	Classes / Object prop. / Data prop.	73 / 17 / 11
Operation research ontology (ORO)	Operation research problems	Base line for representing objective functions and sets of constraints based on mathematical equations	Classes / Object prop. / Data prop.	72 / 52 / 10

The ontological framework has the function to connect the ERP systems to the analytical model responsible for decision-support as explained in subsection 2.2. In this case, the scheduling of pharmaceutical batch plants is considered according to the model presented in section 2.3.

2.1 ANSI-ISA standards

In general terms a standard is a set of characteristics or quantities that describes features of a product, process, service, interface or material. ISA standards help automation professionals to streamline processes and improve industry safety, efficiency, and profitability. ISA standards cover a wide range of concepts of importance to instrumentation and automation at manufacturing. These standards have committees for symbols and nomenclature used within the industry, safety standards for equipment in non-hazardous and hazardous environments, communications standards to permit interoperability equipment availability, and additional committees for many other technical issues of importance. Finally, the standards can be used to provide practical application of expert knowledge reducing downtimes and enhancing operability. Table 2 shows the interaction between ISA-95 for enterprise-to-control system integration and ISA-88 for control system integration in the enterprise control system.

Table 2. Main features of ISA 95/88 standards

Characteristic	ISA-88	ISA-95
Industrial scope	Mainly Batch and Semicontinuous manufacturing.	Spans all types of manufacturing.
Orientation	Physical work execution for Batch and other types of manufacturing.	Definition of work flow and information exchange for manufacturing operations management.
Primary areas of concern	Addresses a lower level, directing, controlling and coordinating the people and equipment that carry out the physical transformation of raw materials into final or intermediate products.	In the way most people describe a manufacturing enterprise, addresses business functionalities and applications at a level below enterprise business systems but above the physical manufacturing equipment.
Manufacturing management functions	Acknowledges but does not directly address manufacturing management functions.	Flexible structure of manufacturing management functions that interacts with business requirements.
Process control functions	Well-defined equipment-oriented process control structure and function hierarchies extending to the pieces of the manufacturing equipment itself.	Stops short after directly addressing most traditional process control activities.

In practice, scheduling information may be maintained in a hierarchy of definitions, corresponding to enterprise-wide routing information, site-wide routing information, and area-wide routing information. Planning and costing information may be combined in integrated ERP systems. Either way, the information needed for costing must be synchronized with the information used in production and the information used in scheduling. ISA standards integrate this information by means the recipe model, which is accepted as the preferred method for implementing flexible manufacturing within the manufacturing industries. A recipe is an entity that contains the minimum set of information that uniquely defines the manufacturing requirements for a specific product. Recipes also provide a way to describe products and how those products are produced. The information contained in these recipes must often be mirrored in other information sets required to operate a manufacturing enterprise. ISA 88 standard defines and makes use of four types of recipes: general recipe, site recipe, master recipe, and control recipe. In general, ISA recipe model provides a common object model for the entire business; allowing easier integration with business (ERP) systems, supply chain management (SCM) systems, product life cycle management (PLM) systems, and production execution systems.

2.2 Ontological framework

This section describes the working procedure of the framework. As a first step, the actual state of the process is captured by the instantiation of the Enterprise Ontology Project (EOP) [16] which contains an integrated representation of the enterprise structure, ranging from the supply chain planning to the scheduling function, thus comprising activities related to the operational, tactical and strategic functions. EOP is structured according to ANSI-ISA 88 and 95 standards nomenclature. As a next step, by means of the instantiation of the Ontological Math Representation [17], various mathematical models (mathematical elements) already established or newly designed models are translated into a semantic representation as mathematical expressions in order to capture the mathematical meaning of enterprise domain elements. This model relates existing classes belonging to the enterprise domain ontology to mathematical elements to understand and translate the system abstraction in equations. In this case, the analytical model is related to the scheduling of multipurpose batch plants. Finally, the Operations Research Ontology [15] formalizes and supports the processes of: i) problem abstraction, ii) analytical model building, iii) problem solution, iv) verification, and v) deployment of the best solution; and allows automated representation of the results of this whole process in standardized formats (e.g., the so-called mathematical programming data formats, such as MPS, nl and LP).

The workflow diagram of the proposed framework is illustrated in Fig. 2. The first phase aims at reaching a formal conceptualization of the real system of the studied industry. This step encompasses the standardized semantic description of the system using the Enterprise Ontology Project, and the definition and acquisition of the required dynamic and static data contained in the ERP systems. In this work, the elements directly linked to the scheduling function are included as presented in the ISA-88 standard in a recipe structure according to different levels (physical, procedural and process). The second phase pursues the future formalization of the mathematical equations describing the system abstraction using the ontological mathematical representation (OMR). This phase results in a potential mathematical description of the entire system. The OMR can capture both mathematical expressions already in use and new developments for the system conceptualization. Specifically in this work, the mathematical equations of the scheduling model adopted as described in section 2.3. Thirdly, the structure of optimization model system along with the mathematical semantic model are the basis for instantiating the operations research ontology (ORO) and obtain a semantic decision model for optimization purposes. ORO has the task of designing the structure and the equation system in order to define a certain problem following the operations research guidelines. Finally, the mathematical programming standards are applied to the semantic decision model and the problem is solved to reach the optimal integrated solution, which assists managers in making the decisions to be deployed in the real system.

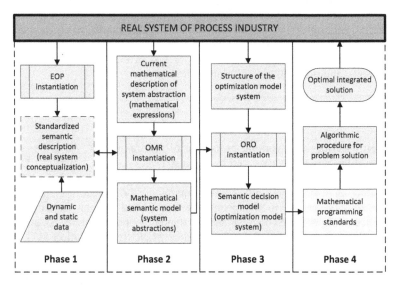

Fig. 2. Flow diagram of the design of problem system construction framework.

2.3 Scheduling problem

The short-term scheduling problem consists of organizing the available human and technological resources in the process plant in order to optimally satisfy customer's demands. Scheduling typically involves decisions on the amount of products to be produced, equipment and resources allocation, production sequence and operations timing, on a weekly or daily basis. A significant amount of models have been proposed in the literature in order to adequately formulate scheduling problems [18,19]. However, each modeling option is only able to cope with a subset of the features, and the choice of the mathematical model has an important impact on computational performance. As a result, the model selection and corresponding problem formulation is a formidable task. Hence, the proposed framework facilitates the creation and application of analytical models for process scheduling relying on the process description available in the ontological model.

This work relies on the mathematical formulation presented in [14] for solving the scheduling problem. It consists of a continuous time state-task-network (STN) formulation including the features of cyclic scheduling, which is often the case for campaign manufacturing in the pharmaceutical industry. This formulation can handle general batch process concepts such as variable batch sizes and processing times, various storage policies or sequence-dependent changeover times. For the sake of brevity, we refer to the aforementioned publication for the explanation of the mathematical model.

3 Case Study

The scheduling problem solved in this work consists of fulfilling the demand for two final products, and the production process consists of four equipment technologies, five production tasks and nine states/materials, specifically three raw materials, two final products and four intermediates (Fig. 3) [14].

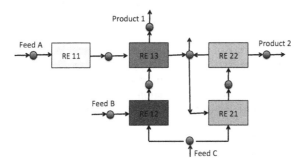

Fig. 3. Diagram of the production process

The problem solution is supported by the knowledge management framework presented in the methodology following the steps of the OR ontology (Fig. 4).

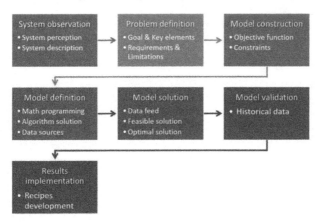

Fig. 4. Main stages of the operations research (OR) problem

As presented in Fig. 4, the processes related to the problem definition, the problem formulation and the solution procedure are represented within the OR Ontology. Next, the performance of the proposed framework is described.

First of all, using the Enterprise Ontology Project, all the features of the production process system have been instantiated. The following phase concerning "Problem definition" aims to capture four key issues: i) goal description, ii) key elements identification, iii) system limitations, and iv) system requirements.

The "Model construction" has been derived from the "Problem definition", which provides the elements for the model creation. Specifically, links among the four different issues identified in the "Problem definition" are established by means of a relation matrix, and the sets, parameters, variables and groups of equations are accordingly derived. Thus, the semantic modeling of the scheduling model is supported by the "Ontological Mathematical Representation", which is also related to the "Enterprise Ontology Project".

The "Data collection" stage relates different data sets and parameters of the problem instantiation to their actual value, which are usually stored in databases. Thus, the "Model solution" selects the solution algorithm according to the specific features of the optimization model, and executes it. Finally, the steps "Model validation" and "Implementation" validate the model using historical data and decision-maker expertise, and an application plan is developed in the real system.

The optimization results of the scheduling are illustrated in Fig. 5. They can be sent back to the ontological model for convenient use by other decision levels, such as the monitoring and control systems. This can be achieved by automatically updating the databases with the resulting optimization data. In fact, several optimization approaches can rely on the same semantic models.

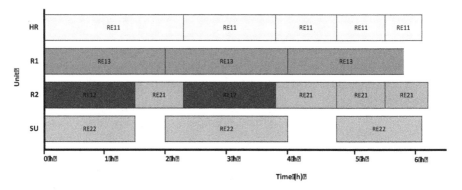

Fig. 5. Gantt chart for the optimal campaign schedule of the process plant (activities represented by boxes, representing start and finishing time at each unit)

4 Conclusions

Pharmaceutical companies strive to remain competitive in a highly challenging environment. Hence, it is of utmost importance to provide adequate tools to support decision-making.

This work presents an ontological framework, which systematically bridges the existing gap between transactional systems and analytical models. Moreover, the presented methodology assists in building computational optimization models for enterprise decision-making in order to facilitate the comprehensive application of enterprise wide optimization throughout the pharmaceutical industry relying on ANSI-ISA standards. The main contribution of this work is the systematic and rigorous representation of the whole decision-making process from the problem conception to implementation and its connection to information systems.

Specifically, the proposed framework comprises the general steps in the area operations research. As a result, a wide range of optimization models could be coupled to existing transactional systems. The case study related to the scheduling problem of a cyclic production plant illustrates information flow as well as the procedure for creation and automation of scheduling models and the full integration and solution of large-scale optimization models.

Future work should focus on the use and interaction of the whole framework at different decision levels in the hierarchical structure of the enterprise.

References

1. Laínez J.M., Schaefer E., Reklaitis G.V.: Challenges and opportunities in enterprise-wide optimization in the pharmaceutical industry. Computers and Chemical Engineering 47, 19-28, (2012)
2. Shah N.: Process industry supply chains: Advances and challenges. Computers and Chemical Engineering 29, 1225-1236, (2005)
3. Elshorbagy, S., Garg, L., Gupta, V.: The Impact of Information Systems on Management Performance in the Pharmaceutical Industry. Journal of Cases on Information Technology, 17(3), 56-73 (2015)
4. Varma, V. A., Reklaitis G. V., Blau G. E., Pekny J. F.: Enterprise wide modeling & optimization - an overview of emerging research challenges and opportunities. Computers and Chemical Engineering 31 (5-6), 692–711, (2007)
5. Fermier, A., McKenzie, P., McWeeney, S., Murphy, T., Schaefer, E.: Stopping Pharma's DRIP (Data-Rich/Information-Poor) problem: A standardized, recipe-based approach adds value and turns data into actionable knowledge. Pharmaceutical Manufacturing, 20–26 (2011)
6. Hailemariam, L., Venkatasubramanian, V.: Purdue ontology for pharmaceutical engineering: Part I. Conceptual framework. Journal of Pharmaceutical Innovation, 5(3), 88–99, (2010)
7. Hailemariam, L., & Venkatasubramanian, V.: Purdue ontology for pharmaceutical engineering: Part II. Applications. Journal of Pharmaceutical Innovation, 5(4), 139–146, (2010)
8. Venkatasubramanian, V., Zhao C., Joglekar G., Jain A., Hailemariam L., Suresh P., Akkisetty P., Morris K., Reklaitis G.: Ontological informatics infrastructure for pharmaceutical product development and manufacturing. Computers and Chemical Engineering 30, 1482–1496 (2006)
9. Morbach J., Yang A., Marquardt W.: OntoCAPE - A large-scale ontology for chemical process engineering, Engineering Applications of Artificial Intelligence 20, 147-161, (2007)
10. Morbach J., Wiesner A., Marquardt W.: OntoCAPE - A (re)usable ontology for computer-aided process engineering. Computers and Chemical Engineering 33, 1546–1556, (2009)

11. Muñoz E., Capon-Garcia E., Moreno-Benito M., Espuña A., Puigjaner L.: Scheduling and control decision-making under an integrated information environment. Computers and Chemical Engineering 35, 774–786, (2011)

12. Muñoz E., Capon-Garcia E., Lainéz J., Espuña A., Puigjaner L.: Ontological framework for the enterprise from a process perspective. In Proceedings of the International Conference on Knowledge Engineering and Ontology Development, SciTePress (Ed.), 538–546 (2011)

13. Muñoz E., Capon-Garcia E., Lainéz J., Espuña A., Puigjaner L.: Supply chain planning and scheduling integration using lagrangian decomposition in a knowledge management environment. Computers and Chemical Engineering 72, 52-67, (2014)

14. Muñoz E., Capon-Garcia E., Lainéz J., Espuña A., Puigjaner L.: Using mathematical knowledge management to support integrated decision-making in the enterprise. Computers and Chemical Engineering 66, 139-150, (2014)

15. Muñoz E., Capon-Garcia E., Lainéz J., Espuña A., Puigjaner L.: Operations research ontology for the integration of analytic methods and transactional data. In proceedings of the 4th International Conference on Software Process Improvement, (2015)

16. Muñoz E., Capon-Garcia E., Lainéz J., Espuña A., Puigjaner L.: Integration of enterprise levels base on an ontological framework. Chemical Engineering Research & Design 91, 1542–1556 (2013)

17. Muñoz E., Capon-Garcia E., Espuña A., Puigjaner L.: Ontological framework for enterprise-wide integrated decision-making at operational level. Computers and Chemical Engineering 42(11), 217–234, (2012)

18. Floudas, C. A, Lin, X.X.: Continuous-time versus discrete-time approaches for scheduling of chemical processes: a review. Comp. Chem. Eng. 28 (11), 2109-2129 (2004)

19. Harjunkoski, I., Maravelias, C.T., Bongers, P., Castro, P.M., Engell, S., Grossmann, I.E., Hooker, J., Méndez, C., Sand, G., Wassick, J.: Scope for industrial applications of production scheduling models and solution methods. Comp. Chem. Eng. 62 (5), 161–193 (2012)

Part III
Software Systems, Applications and Tools

HISMM - Hospital Information System Maturity Model: A Synthesis

João Vidal Carvalho[1], Álvaro Rocha[2], António Abreu[3]

[1] Instituto Politécnico do Porto/ISCAP, S. Mamede de Infesta, Portugal
cajvidal@iscap.ipp.pt

[2] Departamento de Engenharia Informática, Universidade de Coimbra, Coimbra, Portugal
amrocha@dei.uc.pt

[1] Instituto Politécnico do Porto/ISCAP, S. Mamede de Infesta, Portugal
aabreu@iscap.ipp.pt

Abstract. Maturity Models have been introduced, over the last four decades, as guides and references for Information System (IS) management in organizations from different sectors of activity. In the healthcare field, Maturity Models have also been used to deal with the enormous complexity and demand of Hospital Information Systems (HIS). The present paper introduces a maturity model baptized as HISMM, which includes six stages of HIS growth and maturity progression. The HISMM has the peculiarity of congregating a set of key maturity Influence Factors and respective characteristics, enabling not only the assessment of the global maturity of a HIS but also the individual maturity of its different dimensions.

Keywords: Stages of Growth, Maturity Models, Information Systems, Management, Health.

1. Introduction

The confluence between increased computer acceleration capabilities, the reach and expansion of the Internet and the growing ability to capture and leverage knowledge in a digital format, are mainly responsible for the technological revolution we are nowadays experiencing. Our information society can similarly impact healthcare services, changing the relationship between patients and healthcare professionals, that is, providing significant opportunities for healthcare professionals to deliver technology-effective healthcare services to their customers and offering the latter ways to access all the information they need. However, healthcare systems all over the world are undergoing a considerable pressure to reduce continuously rising costs while simultaneously maintaining, or even improving, the quality of healthcare

© Springer International Publishing AG 2017   189
J. Mejia et al. (eds.), *Trends and Applications in Software Engineering*, Advances
in Intelligent Systems and Computing 537, DOI 10.1007/978-3-319-48523-2_18

services [1]. Collateral factors, such as demographic changes, the lack of qualified healthcare professionals, the expectations and the demands from patients, local administrators, and health insurance companies, hinder the attainment of this goal [2]. It is strongly expected that a broader adoption of IS and ICT (Information and Communications Technology) s in the health field will contribute decisively to reduce costs and improve quality [3]. However, evidence shows that implementing IS/ICTs without considering the underlying strategic and organizational structures and processes will not necessarily produce the expected benefits [4]. Indeed, ICTs are generally perceived as holding enormous potential to improve healthcare systems and several examples around the world prove this point. Unfortunately, some cases have also led to disappointment and scepticism [5]. Several studies highlight the importance of finding suitable models that improve, measure and assess the success rate of healthcare systems related projects [5]. Maturity models are a perfect match.

The present paper introduces a Maturity Model baptized as HISMM, which includes six maturity growth and progression stages for a HIS. In the next section maturity models are explained and the generic characteristics of this management tool for organizations are set forth. The following section justifies the need to develop a healthcare related maturity model, as the limitations of existing models are exposed. Finally, the section preceding the conclusion explains the HISMM maturity model as well as the activities that supported its development in the course of our investigation work.

2.   State of the Art

Information Systems Management (ISM) is the activity responsible for the tasks of an organization pertaining to Information Management, Information Systems and the adoption of ICT [6]. The maturity of this activity is a key factor for the success of organizations, to the extent that an IS is fundamental for their survival, competitive edge and success. In this context, there are several instruments that help the ISM achieve an enhanced maturity, namely the so called Maturity Models. Indeed, maturity models provide organization managers an important model for the identification of the maturity stage of an IS in order to plan and implement actions that will allow them to move towards an enhanced maturity stage, and thus achieve the proposed goals [7]. Maturity Models can be perceived as conceptual models, comprised by discreet stages that are used to identify "anticipated, typical, logical or desired evolution paths towards maturity" [8]. We observe that these models have been used in multiple areas to describe a wide variety of phenomena [9-12].

Maturity Models are sustained by the principle that people, organizations, functional areas, processes, etc., evolve, towards an enhanced maturity and following a development or growth process, which covers a number of different stages [7]. That is, Maturity Models are based on the theory of cyclic stages of growth, where the changes observed in an IS over the course of time occur in a sequential and predictable mode, covering a certain number of cumulative and hierarchically sequential stages, which can be described and linked to a specific level of maturity [7, 13-15]. In the same sense, Caralli & Knight [16] argue that maturity models provide organizations with a tool to address their problems and challenges in a structured way, offering both a reference point to evaluate their capabilities and a guide to improve them.

Over the last four decades several maturity models have been proposed, with differences as to the number of stages, influencing factors and intervention areas (Table 1). Each one of these factors identifies the characteristics that typify the focus of each maturity stage, that is, these factors work as reference descriptors or variables to characterize each stage and provide the necessary criteria to achieve a specific maturity stage [8].

Table 1: A maturity model structure [7]

Factors	Stage 1	Stage 2	Stage ...	Stage N
Factor 1	*Characteristic 1* *Characteristic ...* *Characteristic N*	*Characteristic 1* *Characteristic ...* *Characteristic N*	*Characteristic 1* *Characteristic ...* *Characteristic N*	*Characteristic 1* *Characteristic ...* *Characteristic N*
Factor 2	*Characteristic 1* *Characteristic ...* *Characteristic N*	*Characteristic 1* *Characteristic ...* *Characteristic N*	*Characteristic 1* *Characteristic ...* *Characteristic N*	*Characteristic 1* *Characteristic ...* *Characteristic N*
Factor 3	*Characteristic 1* *Characteristic ...* *Characteristic N*	*Characteristic 1* *Characteristic ...* *Characteristic N*	*Characteristic 1* *Characteristic ...* *Characteristic N*	*Characteristic 1* *Characteristic ...* *Characteristic N*
Factor ...	*Characteristic 1* *Characteristic ...* *Characteristic N*	*Characteristic 1* *Characteristic ...* *Characteristic N*	*Characteristic 1* *Characteristic ...* *Characteristic N*	*Characteristic 1* *Characteristic ...* *Characteristic N*
Factor N	*Characteristic 1* *Characteristic ...* *Characteristic N*	*Characteristic 1* *Characteristic ...* *Characteristic N*	*Characteristic 1* *Characteristic ...* *Characteristic N*	*Characteristic 1* *Characteristic ...* *Characteristic N*

2.1 Evolution of Maturity Models in IS Management

Richard Nolan is considered the mentor of the IS maturity approach. Indeed, after studying/researching the use of IS in the biggest US organizations, Nolan proposed a maturity model that initially included 4 stages [14]. Later, with a view to improve his first proposal, Nolan included two additional stages to the initial model [17]. In this second version, Nolan suggests that organizations start slowly in the Initiation stage, followed by a rapid spread in the use of ITs during the Contagion stage. Subsequently, the need for Control emerges, and this stage is followed by the Integration of different technological solutions. Data Management allows for development without increasing IS related costs and, finally, constant growth promotes the achievement of Maturity.

Although this approach to the maturity models developed by Nolan, has been recognized as significantly ground breaking, it also raised a lot of debate and controversy within the scientific community. Several researchers have published studies that, on the one hand, validated and, on the other hand, expanded the model proposed by Nolan. Indeed, resulting from the investigations in this field several researchers have proposed new models (e.g.: [18-22]).

Amongst these new models proposed after the initial approach developed by Nolan, the most widely accepted, detailed and comprehensive is the Revised Model of Galliers and Sutherland [7, 15]. This model provides an improved perspective of how an organization plans, develops, uses and organizes an IS and offers suggestions towards an enhanced maturity stage. This model involves six stages of maturity and assumes that an organization can occupy different maturity stages in any given moment and be conditioned by influencing factors. Moreover, it presents the characteristics of the stages aligned with modern network organizations and offers a data collection tool to evaluate maturity [7].

More recently, after the model proposed by Galliers & Sutherland [22], other models have been resealed (e.g.: [9, 23-25]), including a new Nolan model with nine stages of maturity [26], developed as an answer to the technological evolution in the IS field and its management. As to the field of IS Management, another solid example of a Maturity Model is the model developed by de Khandelwal & Ferguson [25], proposing nine stages of maturity and combining stages theory with Critical Success Factors. Notwithstanding, the model proposed by Galliers & Sutherland [22] is still perceived as the most complete and updated in IS management [15].

Additionally, these maturity models are still being used and implemented in multiple types of organizations and to different areas inside them. Mutafelija & Stromberg [27] refer that the concept of maturity has been applied to more than 150 areas inside IS. In fact, there are several examples of maturity models focused on different organization and IS areas, namely the maturity model for Intranet implementation, by Damsgaard & Scheepers [28]; the maturity model for Enterprise Resource Planning (ERP) systems by Holland & Light [29]; and the CMMI maturity model for the software development process [30]. We can also mention maturity models for fields such as Software Management [31], Business Management [32], Project Management [33, 34], Project Portfolio and Program Management [35], Information Management [36], IS/ICT Management [37], e-Business [21, 38-40], e-Learning [41], Knowledge Management [42, 43], Business Process Management (BPM) [44], Enterprise Architecture [45], etc.

2.2 Maturity Models for HIS Management

Health related organizations, and more specifically Hospital IS Management organizations, are also increasingly adopting maturity models. This use is connected to a growing provision of health care services based on electronic systems, supported by enhanced computer capacity and an increased ability to seize and share knowledge in a digital format. It is widely agreed that ISs offer significant opportunities for health care providers and health provision in general, as well as access to information required by users [15].

In this context, some maturity models emerged, namely the Quintegra Maturity Model for electronic Healthcare [46], proposed as a model that goes beyond the limits of an organization, incorporating every service linked to the medical process applied to each health care provider in each maturity stage. Another example of a maturity model in the health field is the HIMSS Maturity Model for Electronic Medical Record, which identifies different maturity stages in the Electronic Medical Record (EMR) of hospitals [47, 48]. IDC Health Insights has also developed a maturity model which describes the five stages of development for HIS. This maturity model has been used all over the world by IDC, both to evaluate the maturity of IS in hospitals and to compare maturity average differences between regions and countries in different continents [49]. To these models we can add the Maturity Model for Electronic Patient Record, directed to the system that manages every patient related information, that is, a system that manages the EPR (Electronic Patient Record) [50] and the maturity model for PACS[1] by Wetering & Batenburg [51].

National health services from several countries have also started to develop and adopt Maturity Models. That is the case of the model created by the National E-health Transition Authority of Australia [52] and baptized Interoperability Maturity Model (IMM). This model focuses on interoperability associated with technical, informational and organizational capabilities of the different *players* involved in health care services. Another example concerns the Maturity Model of the NHS[2] *Infrastructure Maturity Model* (NIMM) [53]. This is a maturity evaluation model that helps NHS organizations carry out an objective self-assessment in terms of technological infrastructures.

3. HIS Maturity Model: Necessity vs Opportunity

Healthcare and governmental organizations are starting to realize that a certain inability to properly manage health processes is directly connected to limitations in technological infrastructures and to the ineffectiveness of their management [46, 54]. Looking at the healthcare context, the weight and significance of technological transition related problems becomes clear [46]. Moreover, the complexity of IS/ICT operations has grown in order to answer the demands of this sector. This increase in complexity, on the other hand, led to the introduction of multiple new business

---

[1] Picture Archiving and Communication System
[2] National Healthcare Service - UK

integration systems, processes and approaches, as well as the emergence of new companies offering their services in the field. Consequently, immature products and services are being implemented by HIS undergoing changes and demanding, more than ever, a level of performance and effectiveness that answers their actual needs. In this context, it is difficult to know if we are doing a good job in managing these changes and monitoring progress on an ongoing basis. Besides, managing interactions between systems and processes that are constantly evolving is a difficult task, as is managing the impact of low interoperability, safety, reliability, efficiency and effectiveness processes. The benefits of modern technologies in the healthcare field, sustained by improved methods and tools, cannot be harvested through undisciplined and chaotic processes [55]. That is why we believe HIS management must be supported by the implementation of maturity models.

As in most IS fields, several maturity models have been proposed for the healthcare field. Yet, although the specificities of these models set them apart from the ones used in other fields, they remain in an early development stage [15, 56]. In the course of our investigation work we observed that healthcare models are poorly detailed, fail to offer maturity assessment tools and do not structure their maturity stages according to influence factors. Moreover, in our literature review [57], as well as in complementary studies, we noticed the absence, as far as we could tell, of a tolerably encompassing healthcare model which assesses the maturity of HIS in its different components. Indeed, a thorough analysis of scientific papers, manuals, reports and websites addressing healthcare related maturity models revealed an absence of maturity models whose dimensions or influence factors are taken into account with different relative weights according to relevance. Faced with these limitations, we identified the opportunity to develop a sufficiently encompassing maturity model incorporating the main HIS Influencing Factors.

4.   HISMM - Hospital Information System Maturity Model

In order to develop a new HIS Maturity Model, we defined a set of activities in compliance with established research methodologies and considered most suitable for this type of project (see Figure 1). Based on our literature review, the activities included a review of key concepts and aspects of Maturity Models, namely a state of the art review on healthcare IS Maturity Models, as well as the identification and definition of Influence Factors adopted in these Models. The Design Science Research (DSR) methodology, in its turn, supported the activities connected with the identification of the main Influence Factors and the subsequent proposal and validation of the new model. Based on a survey, carried out via questionnaire to 46 HIS experts, we identified the main Influence Factors and developed the first version of our HISMM. Subsequently, the DSR supported our Model validation through a series of interviews carried out with selected group of HIS Managers.

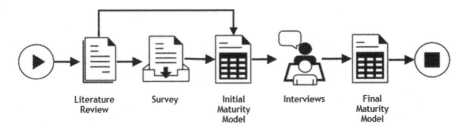

Figure 1. Activities carried out for the development of the HISMM

The HISMM displays a conventional Maturity Model structure, that is, a matrix comprised by different maturity stages and 6 Influence Factors identified as the most relevant for healthcare IS [58]. As seen in Table 2, where our HISMM is briefly described, each factor identifies the features that typify the focus of each maturity stage. These factors emerge as reference descriptors or variables that characterize each stage and determine the necessary criteria to reach a specific maturity stage.

In other words, the HISMM architecture comprehends "levels" (or stages) on an evolutionary scale with measurable transitions between levels. Each level is defined by a set of attributes, and when an HIS reveals such attributes, the corresponding level and the capabilities it embodies have been achieved. With measurable transition states between levels, Hospitals can use this scale to: (1) define the current maturity stage; (2) determine the next achievable maturity stage; (3) identify the attributes that must be met to achieve a new maturity stage.

5. Conclusion

The present paper reports the reasons underlying the development of an encompassing Maturity Model for the healthcare field. The HISMM was developed to address the complexity of HIS and offer a useful tool for the demanding role of HIS management. To validate the HISMM we interviewed a diversified group of IS managers from Portuguese Hospitals. The results of this investigation work have been both encouraging and promising, revealing a high level of acceptance amongst the interviewed managers. This early acceptance pushes us towards the development of a new Model stage, focused on the development of an automatic HIS maturity assessment tool.

Table 2. HISMM Maturity Model

	STAGE I "Ad hocracy"	STAGE II Starting the foundations	STAGE III Centralized dictatorship	STAGE IV Democratic cooperation	STAGE V Entrepreneurial opportunity	STAGE VI Integrated relationships
DATA ANALYSIS	• Isolated and fragmented data analysis solutions • Heavy and complex production of internal and external reports • Data integrity issues • Inability to handle large volumes and variety of data • Problems when collecting data from different systems • Lack of analytical and IT resources • Use of Spreadsheets and local DB	• Key data collection and integration • Centralized data repositories • Automated production of internal reports • Automated production of daily metrics available in BI platforms • Daily productivity is automatically estimated and delivered to managers • Ability to drilldown from a summary to the particular conditions of the patient	• Efficient and consistent report production and adaptability to changing requirements • Decreased variability in healthcare processes and increased focus on internal optimization and waste reduction • Senior managers monitor productivity in terms of staff and combination of skills • Department managers monitor daily productivity results in their dashboards	• Patient care is adjusted based on metrics • Final users have started to incorporate analytical patient data, including Big Data in operations and daily tasks • Costs and quality are monitored via organization performance dashboards • Financial and clinical patient data form a competitive advantage to increase profit	• Organizational processes for intervention are supported by predictive risk models • Clinical risk intervention, modelling and predictive analysis • Full integration of service line data in the strategic planning process • Existence of an Analytics Ecosystem that supports innovation and data exploitation • Clinical outcomes screened with datawarehouses and big data • Alarm management or clinical data intelligence production	• Adoption of personalized medicine and prospective analysis • Patient care adjustment based on population results and genetic data • All valuable data is available for analysis and exploration • Real time data is used in critical activities, such as patient care • Internal and external data sources to improve and optimize costs and quality • Permanent Data Analysis mentality and culture
STRATEGY	• There is no global strategy for IS/IT • There is no formal strategy • Ad-hoc strategies adopted by different IS subareas to answer isolated problems and needs • IT Governance processes are not enforced and the organization does not recognize their need	• Development plans in silos and static structures • Lack of understanding of how to achieve success • The impact of high-level strategies and goals is not mapped • Strategic planning has little impact on day-to-day operations, budgets and resources • Individuals are left on their own to interpret goals, strategies and priorities • IT Governance processes are casual and uncoordinated	• Plans are shared between silos • Different plans with a shared impact are aligned • Low prioritization between groups for high-level projects, goals and plans • There is a measuring tool (although minimal) to assess success and/or impacts • Formal strategy with technology-centric tendency • IT Governance processes follow a regular path	• Strategic plans share a common format • Strategic plans are shared with other strategic initiatives • Available metrics measure the impact of high-level goals in each program • Projects are prioritized based on impact and alignment with established goals • Increasingly inclusive planning for all groups, plans and strategies • IT Governance processes are documented and reported	• Specific group reviews goals and measures progress • Strategic goals become managed programs • The strategy is regularly reviewed and updated • Funding processes aligned to support strategic goals • Faster and more efficient planning and impact analysis • Evolution strategies based on new opportunities and developments in the sector • IT Governance processes are monitored and measured	• Plans are agile and interactive • Plan change impacts are understood and shared with other plans • Projects and costs are measured against strategic goals • Metrics support decision making processes connected with goals and forms of achieving success • Strategic review involves all stakeholders for more comprehensive initiatives • Best IT Governance practices are followed and automated
PEOPLE	• Inconsistency when performing existing practices • Lack of responsibility and capacity of managers/staff • Practices based on customs and habits • Teams lack emotional involvement towards the relevance of usability • Individualist attitude by ICT professionals	• Enforcement of a basic set of people management practices • Adoption of communication and coordination procedures • Introduction of performance management • People receive training • Existence of continuing training plans by type of user • The IS/IT team may receive some training in usability, although obtained during work rather than a formal process	• Development of an infrastructure to increase workforce capacity • Analysis and development of skills • Human resource planning • Recognition of usability value • Internal awareness programs towards usability	• Previously implemented work practices are now standardized and adjusted • Development of careers, work groups and practices based on skills • Participatory culture • Integration of skills at work • Small team with usability related responsibilities • Formal training to expand usability skills • Sharing expert staff with other health units	• Autonomous work groups • Quantitative management of performance and measured practices • Management of organizational capability • Guidance and counselling • All usability benchmarks are implemented, including the existence of a team focused on user experience • Staff is trained and knows how to apply the best practices when developing assessment systems for internal and external use • Healthcare professionals must participate in the definition/design of their clinical pathways	• Continuous improvement of individual and workgroup skills • Workgroup is aligned with organizational capability/performance • Continuous human resource innovation • Business benefits are understood, usability is completely acknowledged, and the results are strategically used by the organization • Ongoing integrated development team training • Users are trained/encouraged to learn new skills

Table 2. HISMM Maturity Model (continued)

	STAGE I "Ad hocracy"	STAGE II Starting the foundations	STAGE III Centralized dictatorship	STAGE IV Democratic cooperation	STAGE V Entrepreneurial opportunity	STAGE VI Integrated relationships
EHR / HAVING MEDICAL RECORD	• Patient clinical data is administrative only • Independent client management and departmental systems • Primary records and clinical images in microfilm or paper • Requires access to paper based systems, because not all repositories are electronic • Relies on statistical formats • Content is kept in separate repositories • There is no Master Patient Index system	• Integrated clinical diagnosis and treatment support • Integrated use of master patient index with departmental systems to organize contents • Early PACS and ERP integration • DICOM images are accessed via separate repositories • Basic scanning of medical records in selected areas only • Record management for physical content only • Electronic integration with administrative systems • Administrative capabilities in resource management, electronic discharge submission, treatment schedule submission and electronic payments	• Clinical activity support • Clinical documentation includes electronic clinical order, report results, prescriptions, multiprofessional care • PACS available outside Radiology/ Medical record recovery (EMR, ECM, DICOM) through portals • Limited EMR and DICOM integration with heavy reliance on unstructured content • Limited electronic record management • EMR with limited interoperability • CIS, LIS, RIS, PACS and medication/pharmacy management systems are implemented	• Adoption of Clinical Knowledge and decision support • Electronic access to guidelines, rules alerts and support systems • Closed circuit medication administration • Large scale PACS dissemination and communication • Internal portals used to access repositories with relevant contents, such as EMR and PACS • EMR connection with Automated ID, with barcodes and OCR for image caption • Static forms replaced by e-forms • Integrated CPOE with billing system	• Medical documents based on structured templates • Outpatient and inpatient regimes • PACS process innovation • Complete PACS and Patient Medical Record integration • Integration of specialized medical modules	• Fully electronic medical records for all areas • Complete recovery of medical records through an EMR based portal • Adoption of mobile telemedicine and wireless access to clinical data • Patient Records becomed a collaborative tool • Complete point-of-care data (tablet, voice or workstation) • Management of audit requests, incidents and investigations • Content organized to support result-based analyses • Use of BPM for cross-functional processes • Medical record use by several healthcare providers
IT SECURITY	• Absence of policies that ensure IT/IS security • Investment in security systems is not a priority • The impact of vulnerabilities is not evaluated	• Early stage of conformity • Lack of policies and procedures defined to protect the organization • No contingency plan in crisis situations • Reactive and unplanned security control • Goals change in response to attacks with the implementation of some type of protection	• Applications and network security is implemented • Changes are not managed in a centralized way and security requests are performed and hoc • Goals focus on business activities of the organization and the protection of central systems • Systems are falsely perceived as being protected • unique credentials to portal access	• Security awareness programs are adopted only for key-resources • IT security procedures are formally defined • Responsibility for IT security is assigned but execution is inconsistent • Able to perform some penetration and detection tests • Closely monitored and mandatory access controls	• Centralized management of security related issues and policies • Users are reliable although system interaction is perceived as a vulnerability • No ad hoc changes • Implementation of central configuration models, from which all settings are derived • Security policies and procedures are in force • Identity management in in/out professionals	• Formal policies and procedures to prevent, detect and correct security problems • Corporate governance aligned with security needs • Internal audit policies with published results and implemented actions • Identification of security issues,/ incidents is systematically monitored • Notification system for security incidents • E-mail filters intrusion detection systems are used
SYSTEMS AND IT INFRASTRUCTURE	• Uncoordinated and unconnected systems with limited applications • LAN infrastructure • Key financial and administrative systems are implemented • Infrastructural management is manual, unarticulated and ad hoc • At this level, the IT focuses on downtime avoidance • Lack of monitoring causes reactive and ad hoc procedures • Unpredictable service performance • Lack of interoperability awareness and supporting processes • Usability focuses on products and processes rather than on people	• Internet based infrastructure with HIPAA • Manual yet coordinated infrastructure management • Knowledge storage in silos • Services are managed and predictable • The organization is focused on obtaining infrastructure control • First interoperability solutions in the clinical/administrative area • Product and service usability focused on users • Sporadic incursion in usability practices with limited resources	• Communication infrastructure based on Secure HL7 • Infrastructure for collaboration and knowledge sharing • Reactive yet becoming proactive • Stable IT infrastructure • The organization recognizes the importance of adopting norms and best practices	• Cooperation infrastructure with medical communities • Adoption of electronic prescriptions • Implementation of international coding of diseases, alerts/ contraindications for educational purposes • Nursing documentation system incorporation • Management of the emergency and cardiology departments • Interoperability guidelines defined for healthcare norms, services, policies, processes and legal compliance	• Fully connected and paper-free infrastructure-SaaS Model • Physician portal and patient portal • Wireless infrastructure • Available processing data tools for research purposes • Consolidated infrastructure level with OaaS Model and RaaS Model • Knowledge sharing and collaboration inside the team • Proactive infrastructure and continuous service improvement • Interoperability assessment	• Infrastructure in a regional/national network that connects all service providers • Aggregate data from all hospitals and regions enable governmental healthcare planning initiatives • Remote patient monitoring and telemedicine • Continuously improving • Interoperability capability based on monitored process feedback • Focused on becoming a catalyst for innovation • Infrastructure for knowledge sharing and business collaborations, internal and external • IT/IS and healthcare stakeholders work as a team

References

1. Fitterer, R. and P. Rohner, *Towards assessing the networkability of health care providers: a maturity model approach.* Inf Syst E-Bus Manage, 2010. 8: p. 309-333.
2. Ahtonen, A., *Healthy and active ageing: turning the 'silver' economy into gold.* European Policy Centre, Europe's Political Economy - Coalition for Health, Ethics and Society (CHES), 2012.
3. Khoumbati, K., M. Themistocleous, and Z. Irani, *Evaluating the adoption of enterprise application integration in health-care organizations.* Journal of Management Information Systems, 2006. 22(4): p. 69-108.
4. Mettler, T., *Transformation of the hospital supply chain: how to measure the maturity of supplier relationship management systems in hospitals?* . International Journal of Healthcare Information Systems and Informatics, 2011. 6(2): p. 1-13.
5. Van Dyk, L. and C.S.L. Schutte, *The Telemedicine Service Maturity Model: A Framework for the Measurement and Improvement of Telemedicine Services.* Intech: open science/open minds Telemedicine, 2013(Chapter 10).
6. Amaral, L. and J. Varajão, *Planeamento de Sistemas de Informação.* 4th Edition. FCA, 2007.
7. Rocha, Á. and J. Vasconcelos, *Os Modelos de Maturidade na Gestão de Sistemas de Informação.* Revista da Faculdade de Ciência e Tecnologia da Universidade Fernando Pessoa., 2004. 1: p. 93-107.
8. Becker, J., R. Knackstedt, and J. Pöppelbuß, *Developing Maturity Models for IT Management – A Procedure Model and its Application.* Business & Information Systems Engineering, 2009. 1(3): p. 213-222.
9. King, W.R. and T.S.H. Teo, *Integration Between Business Planning and Information Systems Planning: Validating a Stage Hypothesis,* . Decision Sciences, 1997. 28(2): p. 279-307.
10. de Bruin, T., et al., *Understanding the Main Phases of Developing a Maturity Assessment Model.* 16th Australasian Conference on Information Systems (ACIS 2005), 2005.
11. Mettler, T. and P. Rohner, *Situational Maturity Models as Instrumental Artifacts for Organizational Design.* DESRIST09, Malvern, PA, USA, 2009.
12. Kohlegger, M., R. Maier, and S. Thalmann, *Understanding maturity models results of a structured content analysis.* Proceedings of I-KNOW I-SEMANTICS, Graz, Austria, 2009: p. 51-61.
13. Bhidé, A.V., *The Origin and Evolution of New Businesses.* Oxford University Press, New York, NY, USA, 2000.
14. Gibson, C. and R. Nolan, *Managing the Four stages of EDP Growth.* Harvard Business Review, 1974. 1: p. 76-88.
15. Rocha, Á., *Evolution of Information Systems and Technologies Maturity in Healthcare.* International Journal of Healthcare Information Systems and Informatics, 2011. 6(2): p. 28-36.
16. Caralli, R. and M. Knight, *Maturity Models 101: A Primer for Applying Maturity Models to Smart Grid Security, Resilience, and Interoperability.* Software Engineering Institute, Carnegie Mellon University, 2012.
17. Nolan, R., *Managing the crisis in data processing.* Harvard Business Review, 1979. 57(2): p. 115-126.
18. McKenney, J.L. and F.W. McFarlan, *The Information Archipelago - Maps and Bridges.* Harvard Business Review, 1982. 60(5): p. 109-119.
19. King, J. and K. Kraemer, *Evolution and organizational information systems: An assessment of Nolan's stage model.* Communications of de ACM, 1984. 27(5): p. 466-475.
20. Huff, S.L., M.C. Munro, and B.H. Martin, *Growth Stages of End-User Computing.* Communications of the ACM, 1998. 31(5): p. 542-550.

21. Earl, M.J., *Management Strategies for Information Technologies.* Upper Saddle River NJ: Prentice Hall., 1989.
22. Galliers, R.D. and A.R. Sutherland, *Information systems management and strategy formulation: the 'stages of growth' model revised.* Journal of Information Systems, 1991. 1(2): p. 89-114.
23. Auer, T., *Beyond IS Implemention: A Skill-Based Aproach to IS Use.* 3rd European Conference on Information Systems, Athens, Greece., 1995.
24. Mutsaers, E., H. Zee, and H. Giertz, *The Evolution of Information Technology.* Information Management & Computer Security, 1998. 6(3): p. 115-126.
25. Khandelwal, V. and J. Ferguson, *Critical Success Factors (CSFs) and the Growth of IT in Selected Geographic Regions.* Proceedings of 32nd Hawaii International Conference on Systems Sciences (HICSS-32), USA., 1999.
26. Nolan, R. and W. Koot, *Nolan Stages Theory Today: A framework for senior and IT management to manage information technology,* . Holland Management Review, 1992. 31: p. 1-24.
27. Mutafelija, B. and H. Stromberg, *Systematic process improvement using ISO 9001:2000 and CMMI.* Boston: Artech House, 2003.
28. Damsgaard, J. and R. Scheepers, *Managing the crises in intranet implementation: a stage model,* . Information Systems Journal, 2000. 10(2): p. 131-149.
29. Holland, C. and B. Light, *A stage maturity model for enterprise resource planning systems.* The Data Base for Advances in Information Systems, 2001. 32(2): p. 34-45.
30. SEI Software Eng Institute, *CMMI® for Development, Version 1.3, Improving processes for developing better products and services.* Tech. Rep. No. CMU/SEI-2010-TR-033), Carnegie Mellon University, 2010.
31. April, A., A. Abran, and R. Dumke, *Assessment of software maintenance capability: A model and its architecture.* Proceedings of the 8th European Conference on Software Maintenance and Reengineering (CSMR2004), Los Alamitos CA: IEEE Computer Society Press, 2004: p. 243-248.
32. Levin, G. and H. Nutt. *Achieving Excellence in Business Development: The Business Development Capability Maturity Model.* 2005   Sep   2014]; Available from: http://www.maturityresearch.com/novosite/biblio/CMM_Achieving%20Excellence%20in% 20Business%20Development.pdf
33. Kerzner, H., *Using the Project Management Maturity Model: Strategic Planning for Project Management (2nd ed).* New York: John Wiley & Sons, 2005.
34. Brookes, N. and R. Clark, *Using Maturity Models to Improve Project Management Practice.* POMS 20th Annual Conference. Orlando Florida USA., 2009.
35. Murray, A., *Capability Maturity Models - Using P3M3 to Improve Performance.* Outperform: MakingStrategy Reality, 2006. 2.
36. Venkatesh, V., et al., *User acceptance of information technology: Toward a unified view* MIS Quarterly, 2003. 27(3): p. 425-478.
37. Renken, J., *Developing an IS/ICT management capability maturity framework.* . Research conference of the South African Institute for Computer Scientists and Information Technologists (SAICSIT). Stellenbosch, 2004: p. 53-62.
38. Earl, M.J., *Evolving the EBusiness.* Business Strategy Review, 2000. 11.
39. Gardler, R. and N. Mehandjiev, *Supporting Component-Based Software Evolution.* . In Aksit, M., Mezini, M. & Unland, R. (Eds.), Objects, Components, Architectures, Services, and Applications for a Networked World, 2003, Series: Lecture Notes in Computer Science, 2591, Springer Verlag., 2003: p. 103-120.
40. Ludescher, G. and M. Usrey, *Towards an ECMM (E-Commerce Maturity Model).* Proceedings of the First International Research Conference on Organizational Excellence in the Third Millennium. Estes Park: Colorado State University, 2000.

41. Marshall, S., *E-Learning maturity model. Process descriptions.* Victoria University of Wellington, New Zealand, 2007.
42. Berztiss, A.T., *Capability maturity for knowledge management.* DEXA Workshop, IEEE Computer Society, 2002: p. 162-166.
43. Maybury, M.T., *Knowledge Management at the MITRE Corporation.* The MITRE Corporation, 2002.
44. Rosemann, M. and T. deBruin, *Business Process Management Maturity - A Model for Progression.* Proceedings of the 13th ECIS, Regensburg, 2005.
45. Nascio, *NASCIO Enterprise Architecture Maturity Model, Version 1.3.* National Association of State Chief Information Officers, 2003.
46. Sharma, B., *Electronic Healthcare Maturity Model (eHMM): A White Paper.* Quintegra Solutions Limited, 2008.
47. Garets, D. and M. Davis, *Electronic Medical Records vs. Electronic Health Records: Yes, There Is a Difference.* A HIMSS AnalyticsTM White Paper, 2006.
48. HIMSS, *The EMR Adoption Model.* HIMSS Analytics: Innovative Research / Informed Decisions, 2008.
49. Holland, M., L. Dunbrack, and S. Piai, *Healthcare IT Maturity Model: Western European Hospitals - The Leading Countries.* European IT Opportunity: Healthcare Healthcare Provider IT Strategies Health Industry Insights, an IDC Company, 2008.
50. Priestman, W., *ICT Strategy 2007-2011 for The Royal Liverpool and Broadgreen University Hospitals NHS Trust.* Trust Board Meeting 6th November 2007. , 2007. Document Number: V1.4.
51. Wetering, R. and R. Batenburg, *A PACS maturity model: A systematic meta-analytic review on maturation and evolvability of PACS in the hospital enterprise.* International Journal of Medical Informatics, 2009. 78: p. 127-140.
52. NEHTA, *Interoperability Maturity Model: Version 2.0.* National E-Health Transition Authority Ltd, 2007.
53. NHS. *National Infrastructure Maturity Model [Online].* . 2011 Sep 2015]; Available from: http://www.connectingforhealth.nhs.uk/systemsandservices/nimm.
54. Freixo, J. and Á. Rocha, *Arquitetura de Informação de Suporte à Gestão da Qualidade em Unidades Hospitalares* RISTI - Revista Ibérica de Sistemas e Tecnologias de Informação 2014. 14: p. 1-18.
55. Gonçalves, J. and Á. Rocha, *A decision support system for quality of life in head and neck oncology patients.* Head & Neck Oncology, 2012. 4(3): p. 1-9.
56. Mettler, T., *A Design Science Research Perspective on Maturity Models in Information Systems.* University of St. Gallen, St. Gallen, 2009.
57. Carvalho, J.V., A. Rocha, and A. Abreu, *Information Systems and Technologies Maturity Models for Healthcare: a systematic literature review,* in *In: Rocha et al. (Eds.), New Advances in Information Systems and Technologies*2016, Springer Berlin Heidelberg. p. 83-94.
58. Carvalho, J.V., Á. Rocha, and A. Abreu, *Main Influence Factors for Maturity of Hospital Information Systems.* 11ª Conferencia Ibérica de Sistemas y Tecnologías de Información Gran Canaria, España, 2016. 1: p. 1067.

Adoption of the User Profiles Technique in the Open Source Software Development Process

Lucrecia Llerena[1], Nancy Rodríguez[1], John W. Castro[2]
Silvia T. Acuña[1]
[1]Departamento de Ingeniería Informática, Universidad Autónoma de Madrid
Madrid, Spain,
[2]Departamento de Ingeniería Informática y Ciencias de la Computación, Universidad de
Atacama, Copiapó, Chile
[1]{lllerena, nrodriguez}@uteq.edu.ec, [2]john.castro@uda.cl, [1]silvia.acunna@uam.es

Abstract. The growth in the number of non-developer open source software (OSS) application users and the escalating use of these applications have led to the need for and interest in developing usable OSS. OSS communities do not usually know how to apply usability techniques in OSS development projects and are unclear about which techniques to use in each activity of the development process. The aim of our research is to determine the feasibility of applying User Profiles in the QUCS OSS project. To do this, we participated as volunteers in this project. We used the case study research method during technique application and participation in the OSS community. As a result, we identified adverse conditions that were an obstacle to technique application and modified the technique to make it applicable. We conclude that it was not easy to recruit OSS users and developers to participate in usability technique application.

Keywords: open source software, usability techniques, requirements engineering, user analysis, user profiles.

1 Introduction

Open source software (OSS) has spread so swiftly that it now rivals commercial software systems [1]. OSS communities do not as yet enact standard processes capable of ensuring that, considering the characteristics of the OSS community as a whole, the software that they develop has the attributes of good software [2]. An inadequate definition of processes, activities, tasks and techniques within OSS development has led researchers from several areas to gravitate towards this field of research with the aim of correcting this situation [3], [4], [5].

Usability is one of the key quality attributes in software development [6]. In recent years, OSS has come to be an important part of computing [7], [8], [9], [10]. However, several authors have acknowledged that the usability of OSS is poor [11], [12], [13]. In this respect, the empirical study conducted by Raza et al. [14] reports that 60% of respondents (non-developer users) stated that poor usability is the main obstacle to be overcome by OSS applications if users are to migrate away from

J. Mejia et al. (eds.), *Trends and Applications in Software Engineering*, Advances
in Intelligent Systems and Computing 537, DOI 10.1007/978-3-319-48523-2_19

commercial software. On this ground, OSS projects must tackle the usability level and usability-related problems in more depth [13]. On one hand, the human-computer interaction (HCI) field offers usability techniques whose key aim is to build usable software. However, they are applied as part of HCI methods and not within the OSS development process. On the other hand, the OSS development process focuses on source code and thus on the development of functionalities. The OSS development process has a number of features (like functionality-focused development) which prevent many of the HCI usability techniques from being adopted directly [15].

This community has now started to adopt some usability techniques. Most of the techniques taken on board by the community are for evaluating usability [15], whereas it has not adopted many techniques related to requirements analysis and design. Some techniques have been adapted ad hoc for adoption in OSS development projects [15]. This paper addresses the research problem of how to adopt User Profiles. To do this, we previously identified and analysed which obstacles had to be overcome in order to be able to apply this technique in OSS projects. Some OSS projects are beginning to adopt HCI usability techniques. Ferré [16] compiled a list of techniques recognized by HCI and determined the major software engineering (SE) activities in which they are used.

Our research spans two areas: SE and HCI. We use usability techniques as a bridge to communicate these two areas, where our aim is to deploy knowledge of HCI in the SE field and especially in the OSS development process. It is crucial to study the OSS development process in order to ascertain how this community develops software and determine which points should be taken into account when using usability techniques in OSS projects [15]. By adapting usability techniques, we can adopt OSS development process activities according to the standards of systematization current in SE. Therefore, this paper has two goals. Firstly, we intend to adapt the User Profiles usability technique [17] for adoption in the OSS development process. Secondly, we aim to determine the feasibility of adopting this usability technique in real OSS projects, particularly Quite Universal Circuit Simulator (QUCS).

Requirements engineering activities play a very important role in the success or failure of an OSS project, but they are sometimes extremely hard to perform because the OSS user segments are not previously defined. Also, it is by no means straightforward to address all requirements analysis activities due to the particular characteristics of OSS development groups (for example, global geographic distribution of user sites or code-focused world view). On this ground, we consider the User Analysis activity in this paper. Additionally, OSS projects have not adopted many usability techniques related to requirements engineering activities [15].

Some authors claim that the main reasons for the generally poor usability of OSS developments are that OSS developers have tended to develop software for themselves [4], [18] and that the development community is uninformed about who its users are [11], [19]. The User Profiles technique is useful for gathering information about the intended system users [16]. This technique provides details about the prospective system users and provides user-centred guidance for application design [20]. In other words, the technique helps to familiarize developers with their users. On this ground, we selected the User Profiles usability technique for adoption in OSS.

There are several OSS project repositories. One of the most popular is SourceForge.net [21]. This repository classifies OSS projects by categories. We looked at projects with a low level of coding (that is, projects where key functionalities are still being added) that were not overly ambitious and were at the very early stages of development (alpha) in order to select a suitable OSS project for adopting the usability techniques. Additionally, we considered the number of potential users because the existence of a critical mass dictates the real user segments of an application. Other selection criteria considered are a higher activity frequency and rating (that is, the popularity of an application scored by users). We selected the QUCS OSS project. As the project is at the early stages, we can adopt usability techniques related to requirements activities. Therefore, the benefits of applying the techniques will have a bigger impact on the development process and software system usability. We have adapted the techniques based on the integration framework proposed by Castro [15]. We used a case study as the research method for testing the feasibility of our proposal for adopting usability techniques in OSS projects [22]. Consequently, we had to volunteer for the OSS QUCS project and become members of the community.

This paper is organized as follows. Section 2 describes the research method followed to apply the usability technique. Section 3 describes the proposed solution. Section 4 discusses the results. Finally, Section 5 outlines the conclusions and future research.

2 Research Method

As mentioned above, we used a case study as the research method to validate our research [23]. From a case study we learn about the experiences of applying usability techniques adapted to OSS projects. This research method is used when the phenomenon under investigation (in this case, the adoption of a usability technique with adaptations) is examined within its real setting (in this case, an OSS project). OSS projects are an adequate setting for the study conducted in this research because OSS communities are generally uninformed about usability techniques, do not have the resources to test usability and cannot usually count on usability expert involvement [4], [11], [19]. The case study addresses the following research question (RQ): Is it possible to determine whether certain adaptations of usability techniques enable their use in requirements engineering activities for adoption in OSS projects?

This research question refers to the User Profiles technique in the QUCS project. QUCS software development projects have a low level of coding, are not overly ambitious and are at the very early development stages (alpha). Additionally, their user segments have not been previously defined. In this paper, we have created web artefacts to improve communication with members of the OSS communities and to efficiently synchronize the activities required to apply the usability technique. The web artefacts used to test the feasibility of the proposed adaptation of each technique were: wiki and online questionnaires (Google Forms). We used Google Forms to gather and organize information on possible application users. Additionally, a wiki was set up for all users to enter their opinions with a view to fine tuning the questionnaires built using Google Forms. These artefacts constitute a virtual meeting point for OSS users that are geographically distributed all over the world. The aim behind the use of these web artefacts was to independently generate results for each case by systematically generating the key user characteristics. Later, we apply the usability technique in the case study to identify the obstacles to its use. Finally, we discuss how they were adopted in the OSS projects.

3 Proposed Solution

In this section, we briefly describe the User Profiles usability technique applied in an OSS project. Firstly, we specify the characteristics of the selected OSS project (QUCS). Second, we describe the User Profiles technique as prescribed by HCI, followed by the details of the changes made to this technique for application in the OSS project. Finally, we report the results of applying the User Profiles technique.

Case Study Design. Case studies are one of the most popular forms of qualitative empirical research [24]. A case study investigates the phenomenon of interest in its real-world context. The phenomenon of interest for this research is the adoption of the User Profiles technique with adaptations, whereas the real-world context is an OSS project. It is by no means easy to run controlled experiments on OSS because it is troublesome to control all the characteristics of OSS communities (for example, age, availability, expertise, experience, etc.). Not all OSS project team members have the same characteristics. Therefore, it is impossible to minimize the effects of external factors (for example, geographic distribution and time differences) in order to effectively evaluate our case studies. We describe the case study following the guidelines set out by Runeson and Host [22]. According to these guidelines, we divide our research into two parts: an exploratory part and a descriptive part. We start by looking at what happens in a real-world scenario and then we describe what happens when we apply the adapted techniques to improve application usability [22]. We selected QUCS as the OSS project in which to adopt User Profiles. QUCS is a multiplatform application for simulating electronic circuits. The size of this application is 504,526 lines of code, written primarily in C++. It was developed under the GNU/Linux operating system, although it also operates on Windows, Solaris, NetBSD, FreeBSD, MacOS and Cygwin. Additionally, it has a GPL licence, which means that it is free for programmers and users. The application has 208 subscribers (quantity supplied by the project administrator via email). Additionally, 2,222 downloads per week are reported on the project web site.

Changes to the User Profiles Usability Technique. The main aim of HCI usability techniques is to produce usable software. However, they tend to be applied only as part of HCI methods and not in the OSS development process [15]. According to HCI recommendations, the User Profiles technique is capable of defining representative classes of users to drive application design [25]. This technique has been applied by administering structured interviews to members of the OSS community with HCI students under the supervision of a mentor acting as usability experts.

The User Profiles usability technique cannot be applied directly in the OSS development process because this community has characteristics that are out of the ordinary in the HCI world, such as, for example, the global geographic distribution of its members, a code-focused world view, shortage of resources and a culture that may be somewhat alien to interaction developers. Even though usability techniques require conditions with which OSS projects cannot generally comply, the techniques can be adapted to bring them more into line with the idiosyncrasy of OSS projects. In the following, we describe the adaptations that we have analysed with a view to adopting the User Profiles techniques in OSS projects.

When a usability technique has to be applied by a usability expert, we recommend that a developer, HCI student or student group (under the supervision of a mentor) or users applying the technique with the help of a document (for example, a style guide) should stand in. In other cases, when user participation is necessary in order to apply the usability techniques, we suggest that OSS users should participate remotely via, for example, chats, forums, blogs and wikis. Other HCI usability techniques, like User Profiles for example, require not only user participation but also that they meet in person. As a result of our analysis, we propose that meetings be held virtually via chats, forums, wikis or by users giving their opinion remotely via email. In the analysed literature [16], [17], there are different procedures for applying the User Profiles technique. The procedure proposed by Mayhew [17] is a good option because it is the approach that gives the most comprehensive description of what to do and how to apply the technique. According to Mayhew [17], the User Profiles technique is composed of 12 steps. In the following, due to space restrictions, we describe the first three steps of this technique as enacted in the QUCS project application.

During Step 1 (Determine user categories), we consulted the project administrator about possible user categories. Similarly, at a meeting of team members in Step 2 (Determine key user characteristics), we asked their opinion. This was then used to design a questionnaire template. In Step 3 (Design draft questionnaire), we revised and expanded the questionnaire template designed in Step 2. We conducted a similar analysis for each of the 12 steps of the User Profiles technique in order to identify which tasks need to be carried out to adapt and apply the technique. Below, due to space restrictions, we describe how the first six steps of the User Profiles technique were adapted.

In Steps 1 and 2, Determine user categories and characteristics, project developers and users should, according to HCI recommendations, meet in person, a condition that cannot be met due to the characteristics of OSS projects. Therefore, the technique requires modification, a modification consisting in asking the project administrator to give feedback remotely and asynchronously (via email). In Steps 3, 4 and 5, Design, gather feedback and revise draft questionnaire, it is not feasible to hold face-to-face meetings to discuss the structure and content of the draft questionnaire considering that the OSS community members are distributed all over the world. Instead we suggest that the project administrator should revise the questionnaire design via email. Step 6 states that the questionnaire has to be refined at a meeting with the application users. This again is an obstacle because the users are distributed all over the world. To overcome this obstacle, we suggest that users should participate remotely via virtual meetings conducted through a web artefact (like a wiki). Table 1 shows the steps of the User Profiles technique, the identified adverse conditions and the proposed adaptations. There are two main adaptations: (i) users should participate online via email and a wiki, and (ii) a developer, an expert user or an HCI student under the supervision of a mentor should stand in for the usability expert. In this particular case, a group of HCI students under the supervision of a mentor substituted the expert.

Table 1. Summary of the identified adverse conditions and the proposed adaptations for the User Profiles technique.

Technique Steps [17]		Adverse Conditions	Proposed Adaptations
1	Determine user categories	The OSS project developers must meet in person.	Email the project administrator to ask for a description of the intended user categories.
2	Determine key	Several users must meet in	Email users to ask for a description

		person.	the key characteristics of their profile.
	user characteristics	person.	the key characteristics of their profile.
3	Design the draft questionnaire		Email the project administrator several times to ask for revisions of the draft questionnaire.
4	Gather feedback about the draft questionnaire		
5	Revise the questionnaire		
6	Pilot the questionnaire	Users must participate face to face.	Meetings are virtual using a wiki where users can post their opinion on questionnaire design instead.
7	Fine tune the questionnaire according to the results of the pilot questionnaire		
8	Select a user sample	Developers must participate.	Ask the project administrator for the user mailing list.
9	Distribute questionnaires		
10	Design the data input format		
11	Data statistics (enter, summarize and interpret)	A usability expert is required to apply this step of the technique.	Replace the usability expert with: an HCI student or student group (supervised by a mentor).
12	Report data	(This adaptation is a response to the OSS work method rather than a solution to an adverse condition.)	Publish the conclusions of the study in forums and report back to the OSS community via mailing lists.

Case Study Results. We applied the adapted User Profiles technique to the OSS QUCS project, a multiplatform application for simulating electronic circuits. QUCS application users are mostly students that are new to the field of electronics. The first thing that we did was to contact the project administrator who was receptive about applying this technique. The QUCS users showed interest in participating and learning more about the User Profiles technique. We designed a questionnaire based on Mayhew's proposal in order to create the User Profiles [17]. We created the "circuitoselectricoslu" wiki using the PBWorks tool in order to encourage user participation in initial questionnaire design. Then with the comments collected on the wiki and the feedback provided by the administrator we have improved the initial version of the survey. In the same way, with the improved version of the survey we created an online survey to be published in the forum of the SourceForge community. Finally, 14 people completed this survey over a period of 15 days. The online survey used to gather the data is available at the web site[1]. Thanks to the suggested adaptation for the User Profiles technique, we were able to approximate the QUCS

---

[1] http://goo.gl/EPsIz6

application user type: one segment composed of students and another segment composed of professors.

4 Discussion of Results

The obstacles to applying the User Profiles technique in the QUCS project that we identified concerned how to gather the key characteristics of a homogeneous user profile on mainly two grounds. Firstly, the global geographical distribution of users and time differences made it hard to arrange a meeting with users to discuss the structure and content of the questionnaire. Secondly, an expert in the area was required to analyse the collected data. We found a solution to the problems encountered in our study with regard primarily to user participation by searching the social networks for users. In this manner, we improved user communication and participation in our research. Nowadays, social networks are a tool for disseminating scientific knowledge and research results to a global audience. The instruments used in this research were validated by email exchange with the project administrators. The aim of this exchange was to get feedback from administrators. Thanks to this feedback, we managed to identify and correct potential sources of error in these instruments before applying the final version.

In regard to our proposal to substitute a developer, expert user or HCI student under the supervision of a mentor for the usability expert, the expert was replaced in this particular case by a group of HCI students under the supervision of a mentor. Note importantly that these were final-year Master in Information Communications and Technology Research and Innovation (i2-ICT) students who had taken two HCI courses. Additionally, the students were supervised by two expert usability researchers. On this ground, there is no risk of the application of the proposed adaptation of the selected techniques having a negative impact on software quality.

Additionally, not only are the stated problems and solution proposed in this research of interest and importance to OSS development communities; they are also a concern in global software development (GSD) because GSD business and industrial settings share a number of the characteristics of OSS communities. GSD is now a major issue in SE research and practice [26]. Some negative factors that GSD has to tackle are neglect by project managers, member participation and allocated resources influencing organizational success [27]. Against this backdrop, the results of our research can be extrapolated in order to deal with the above negative factors within the GSD field.

Here, we describe some statistics considering the results of the 14 people who responded to the online survey. First, the respondents (about 50%) have a significantly experience higher to 6 years working with circuits. For this reason, it is convenient to use software applications for this purpose. Second, half of respondents have used the application between 2 and 5 years, this confirmed the experience specifically in the use of QUCS. Third, most participants (85.7%) indicated that work up to three hours with QUCS, so we can affirm that the application QUCS plays an important role in the work of its users. Finally, analyzing that 79% of participants are over 25 years and approximately 64.2% of them are professionals in the field of electronics, it is possible to said that the profile of QUCS users are professionals and not students.

5 Limitations of the study

The case study has some limitations to validate their contributions by its eminently qualitative practice [28]. We have identified three limitations. First, the design of the survey through the wiki, only one person provides feedback to improve the design of the survey. The information collected in the survey design was sent to the OSS project administrator to make the necessary corrections. That is, although the instrument to collect data was validated by the project manager, this lacks of more feedback before applying the final online survey (construct validity). Second, age, level of study and experience of using the tool that the users have may have influenced for or against the application of the technique usability. Initially, the credibility of the survey can vary depending on which data are omitted or the meaning of so the question is distorted. It is need to reapply the online survey at different times of the research process for that the analysis of results is not skewed (internal validity). Third, the main limitation of our study is the number of case studies. In our particular case, we transform a technique to be applied to a single OSS project. Therefore, missing further case studies to apply the technique of usability User Profiles in other OSS projects to validate the proposed changes (external validity).

6 Conclusions and Future Research

The goal of this research paper was to evaluate the feasibility of adopting the User Profile HCI usability technique in an OSS project. We adapted this technique for adoption. Through adaptation, we were able to account for some OSS development characteristics that pose an obstacle to the application of the techniques as per HCI recommendations (for example, OSS developers and users are geographically distributed). It is not easy to recruit volunteer users to participate in projects of this type. Users do not usually have much time, and it is hard to get them to participate without an incentive. By applying the User Profiles technique, we managed to define user segments by publicizing the information-gathering instruments on social networks rather than thanks to the participation of OSS community members (contacted via an OSS forum). After analyzing and applying the usability technique, we found that there are adverse conditions such as the number of participant users, biased information supplied by developers, geographical distribution and time differences and OSS community motivation.

We can conclude that the results with regard to OSS user participation were not what we expected for two reasons. Firstly, based on the statistics reported on the application (QUCS) web site, we thought that we would get a lot of users to participate. Secondly, it was hard to make contact with and engage users in this research. Note that OSS community members are all volunteers, and they use their spare time to do this work.

We believe that, in order to improve the integration of usability techniques into OSS projects, the OSS community has to start attaching importance to and raising awareness of the repercussions of the issues addressed by the HCI field on software development. As it needs to adapt HCI techniques for adoption in OSS development projects, the OSS community also has to broaden its view of software development in order to consider usability and not focus exclusively on developing functionalities. In the future, we aim to conduct further case studies to adapt and apply other usability techniques in OSS projects. We will analyse other web artefacts that can be adapted to improve communication in OSS communities (for example, social networks) and gradually raise the awareness of OSS developers about the benefits of applying HCI usability techniques.

Acknowledgments. This research was funded by the Secretariat of Higher Education, Science, Technology and Innovation (SENESCYT) of the Government of Ecuador and Quevedo State Technical University through doctoral scholarships for higher education professors.

References

1.	Schryen, G., Kadura, R.: Open Source Vs. Closed Source Software. In: 2009 ACM Symposium on Applied Computing - SAC '09. pp. 2016–2023. ACM (2009).
2.	Noll, J., Liu, W.-M.: Requirements Elicitation in Open Source Software Development: a Case Study. In: 3rd International Workshop on Emerging Trends in Free/Libre/Open Source Software Research and Development - FLOSS '10. pp. 35–40. ACM (2010).
3.	Madey, G., Freeh, V., Tynan, R.: The Open Source Software Development Phenomenon: An Analysis Based on Social Network Theory. In: Eighth Americas Conference on Information Systems. pp. 1806–1813 (2002).
4.	Nichols, D.M., Twidale, M.B.: The Usability of Open Source Software. First Monday. 8, 21 (2003).
5.	Raza, A., Capretz, L.F., Ahmed, F.: Maintenance Support in Open Source Software Projects. In: Eighth International Conference on Digital Information Management (ICDIM 2013). pp. 391–395. IEEE (2013).
6.	Ferré, X., Juristo, N., Windl, H., Constantine, L.: Usability Engineering-Usability Basics for Software Developers. IEEE Softw. 18, 22–29 (2001).
7.	Hars, A., Ou, S.: Working for Free? – Motivations of Participating in Open Source Projects. In: 34th Hawaii International Conference on System Sciences. pp. 1–9. IEEE (2001).
8.	Mockus, A., Fielding, R.T., Herbsleb, J.D.: Two Case Studies of Open Source Software Development: Apache and Mozilla. ACM Trans. Softw. Eng. Methodol. 11, 309–346 (2002).
9.	O'Mahony, S.: Guarding the Commons: How Community Managed Software Projects Protect their Work. Res. Policy. 32, 1179–1198 (2003).
10.	Scacchi, W.: Understanding Requirements for Open Source Software. In: Lecture Notes in Business Information Processing. pp. 467–494. Springer-Verlag. (2009).
11.	Benson, C., Müller-Prove, M., Mzourek, J.: Professional Usability in Open Source Projects: GNOME, OpenOffice.org, NetBeans. In: CHI'04 Extended Abstract on Human factors in Computing System - CHI EA'04. pp. 1083–1084. ACM (2004).
12.	Çetin, G., Gokturk, M.: A Measurement Based Framework for Assessment of Usability-Centricness of Open Source Software Projects. In: 4th International Conference on Signal Image Technology and Internet Based Systems - SITIS'08. pp. 585–592. IEEE (2008).
13.	Raza, A., Capretz, L.F., Ahmed, F.: Users' Perception of Open Source Usability: An Empirical Study. Eng. Comput. 28, 109–121 (2012).
14.	Raza, A., Capretz, L.F., Ahmed, F.: An Empirical Study of Open Source Software Usability: The Industrial Perspective. Int. J. Open Source Softw. Process. 3, 1–16 (2011).

15. Castro, J.W.: Incorporación de la Usabilidad en el Proceso de Desarrollo Open Source Software., (2014).
16. Ferré, X.: Marco de Integración de la Usabilidad en el Proceso de Desarrollo Software., (2005).
17. Mayhew, D.J.: The Usability Engineering Lifecycle: A Practitioner's Handbook for User Interface Design. Morgan Kaufmann. (1999).
18. Raza, A., Capretz, L.F., Ahmed, F.: An Open Source Usability Maturity Model (OS-UMM). J. Comput. Hum. Behav. 28, 1109–1121 (2012).
19. Nichols, D.M., Twidale, M.B.: Usability Processes in Open Source Projects. Softw. Process Improv. Pract. 11, 149–162 (2006).
20. Cooper, A., Reinmann, R., Cronin, D.: About Face 3.0: The Essentials of Interaction Design. Wiley (2007).
21. SourceForge, https://sourceforge.net/.
22. Runeson, P., Host, M., Rainer, A., Regnell, B.: Case Study Research in Software Engineering: Guidelines and Examples. John Wiley & Sons. (2012).
23. Coutaze, J.: Evaluation Techniques: Exploring the Intersection of HCI and Software Engineering. In: Taylor, R.N. and Coutaz, J. (eds.) Software Engineering and Human-Computer Interaction. pp. 35–48. Springer (1995).
24. Runeson, P., Höst, M.: Guidelines for conducting and reporting case study research in software engineering. J. Empir. Softw. Eng. 14, 131–164 (2009).
25. Hix, D., Hartson, H.R.: Developing User Interfaces: Ensuring Usability Through Product & Process. John Wiley & Sons. (1993).
26. Aurora Vizcaíno, García, F., Piattini1, M.: Visión General del Desarrollo Global de Software. Int. J. Inf. Syst. Softw. Eng. Big Co. 1, 8–22 (2014).
27. Khan, A.A., Keung, J.: Systematic review of success factors and barriers for software process improvement in global software development. IET Softw. 1–11 (2016).
28. Pérez, J.: Dialnet-El Estudio de Casos como Estrategia de Construcción Teórica, http://dialnet.unirioja.es/servlet/articulo?codigo=195459&info=resumen&idioma=SPA, (1999).

The Use of Inverted Index to Information Retrieval: ADD Intelligent in Aviation Case Study

Sodel Vázquez-Reyes, María de León-Sigg, Perla Velasco-Elizondo, Juan Villa-Cisneros, Sandra Briceño-Muro,

Autonomous University of Zacatecas, Software Engineering
Zacatecas, Mexico
vazquezs@uaz.edu.mx, mleonsigg@uaz.edu.mx, pvelasco@uaz.edu.mx, jlvilla@uaz.edu.mx, sgbm0592@gmail.com

Abstract. Nowadays store, index and retrieve information from document collections is a complex but necessary task. For this reason, information retrieval is fundamental to decision-making in companies. The *Be Intelligent* system offers a solution to storing, indexing and retrieval of documents content of ADD Intelligent Aviation company. The system performs searches through natural language expressions, presents the user a list of results containing document name, page, author, date and paragraph with search terms highlighted. The list of documents that meets the search is ordered by the relevance between the expression in natural language and the content of a document.
Be Intelligent system provides support for administration, indexing and retrieval of digital documents that the company uses during inspections of aircraft, reducing time to retrieve information.

Keywords: inverted index, information retrieval, precision, mean reciprocal rank.

1 Introduction

Presently is increasingly evident the urgency to search for information in large repositories that contain information in documents rather than data. This has transformed information retrieval into an important field of research, and in the development of computer systems that must a) be able to process quickly large collections of documents; b) allow flexible search operations, and c) allow classification of recovered information [1]. In this regard, the final purpose of information retrieval systems is to offer mechanisms that allow companies to acquire, produce and transmit, at the lowest cost, data and information with the attributes of quality, precision, and validity, in order to be useful to decision-making [2]. However, it is important to specify that the main function required from a retrieval information system is not to return the information desired by the user, but to indicate which documents are potentially relevant to his need of information, because, in fact, a user of an information retrieval system is interested about some subject, and not in the specific data that satisfy a query. Information retrieval systems deal with text,

© Springer International Publishing AG 2017
J. Mejia et al. (eds.), *Trends and Applications in Software Engineering*, Advances in Intelligent Systems and Computing 537, DOI 10.1007/978-3-319-48523-2_20

generally written in natural language, not well structured, and semantically ambiguous [3]. Consequently, retrieved documents are judged as useful or not useful, and usefulness is judged in terms of degrees of effectiveness, being the standard measure the utility of the retrieved document [4]. As utility is a complex criterion to measure, because it is mainly based on the judgment of someone (user or non user) [5], several metrics are commonly used. Examples of them are *Precision*, *Reciprocal Rank* and *Mean Reciprocal Rank*. *Precision* (1) is defined as the proportion of recovery relevant documents [6]. This metric evaluates the system ability to position first most of relevant documents and measures the percentage of recovery documents that have relevance [6]. Its calculation is obtained with

$$Precision = \frac{number\ of\ relevants\ recovered}{number\ of\ results} \tag{1}$$

On the other hand, *Reciprocal Rank – RR* (2) is used to measure the system ability to retrieve relevant documents in the higher positions in the list of result. It is calculated by the next equation [7]:

$$RR = \frac{1}{rank(i)} \tag{2}$$

where *rank(i)* refers to the position of the document that contains the correct information to query *i*, and *RR* will be *cero* if information searched is not located in any document.

Finally, the *Mean Reciprocal Rank –MRR* (3) is the average or the *RR* values for all of the queries [7]. This metric gives the highest score to documents that are in the first positions of the list of results, because it measures precision and the order of the correct results [7]. This metric is calculated with

$$MRR = \frac{\sum_{i=1}^{|Q|} RR}{|Q|} \tag{3}$$

where *RR* is the *Reciprocal Rank*, shown in (2), and *Q* is the number of results. To facilitate retrieval of documents or parts of documents, indexes are created. Indexes, then, provide users with effective and systematic means for locating documentary units relevant to information needs or requests [8]. There are several index structures, but one of the most popular is inverted index. An inverted index is a mechanism oriented to words formed by two elements: 1) the vocabulary, defined as the set of different terms (words) in texts; and 2) the occurrence lists, defined as the list of documents in which a given term appears [1].

In this document is presented the use of *Be Intelligent*, a retrieval information system for company (*ADD Intelligence in Aviation*). The system uses an inverted index to retrieve information from a document collection stored in an external hard drive. This research first outlines *ADD Intelligence in Aviation* needs of information, describing briefly the use they give to it. This company represents the case study presented in this research. Next follows the description of the use of the inverted index data structure to retrieval information. Later the results of the implementation of the inverted index are presented. The document concludes by discussing the results of the implementation and the conclusions reached.

2 ADD Intelligence in Aviation case study

ADD Intelligence in Aviation is a center of aeronautical engineering specialized in physical inspections on airplanes. Nowadays, *ADD Intelligence Aviation* works with clients distributed in Mexican cities as Toluca, Ciudad de México and Sonora, but also develops projects in Seattle, WA., Santiago de Chile, in Chile, and Quetta, in Pakistan.

When an inspection project begins, engineers of *ADD Intelligence in Aviation*, gather different manuals, inspection templates, and other needed documents to verify an assigned aircraft. These documents are stored in different formats, including **.pdf, *.docx, and *.xls*. The document collection is fed with manuals developed by *ADD Intelligence in Aviation*, inspection templates, and airworthiness directives obtained in web sites of different aviation agencies, such as the International Air Transport Association, the European Aviation Safety Agency, and the American Airlines Maintenance. The document collection is stored in an external hard drive belonging the company. This was a company's internal decision. Access to this hard drive is made through *ADD Intelligence in Aviation* local networks. This situation implies that if an engineer is working outside the reach of local network, a collection of documents is send to him via e-mail, or by using a file-sharing service as Google Drive or Dropbox. Because all of this, it is needed a system to consult the content of documents and confirm if it has been useful to aircraft inspection.

3 Inverted index data structure to information retrieval

It is a well-known fact that building indexes is needed to implement efficient searches. A data structure called inverted index which given a term provides access to the list of documents that contain the term. The inverted index is the list of words and the documents in which they appear.

Most operational information retrieval systems are based on the inverted index data structure. This enables fast access to a list of documents that contain a term along with other information (for example, the weight of the term in each document, the relative position of the term in each document, etc.). In information retrieval, objects to be retrieved are generically called "documents". Given a user query, the retrieval engine uses the inverted index to score documents that contain the query terms. Terms that are considered non-informative, like function words (the, in, of, a, etc.), called stop-words, are often ignored.

Inverted index exploits the fact that given a user query, most information retrieval systems are only interested in scoring a small number of documents that contain some query term. Since all documents are indexed by the terms they contain, the process of generating, building, and storing document representations is called indexing and the resulting inverted files are called the inverted index. Building an inverted index for maintaining any kind of searching system requires to perform a series of steps: fetching the documents, removing the stop words, and finally merge and store the terms to inverted index [1]. This process of index construction or indexing is shown in Fig. 1.

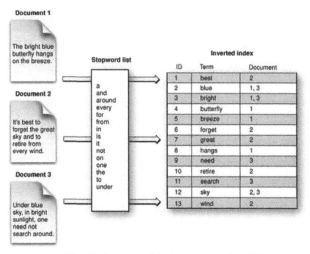

Fig. 1   Process of index construction [9]

4   Experiment design

As discussed in Section 1, there are several metrics to evaluate an information retrieval system. In this section are presented the values obtained for those metrics with the *Be Intelligent* system.

To evaluate *Be Intelligent* system, a collection of several documents was formed. The collection contained 300 documents: 100 of them contained information related to aircraft control, maintenance, and inspections; 115 documents contained information about software engineering, and 85 are documents with guides to research development and information retrieval. Also, two groups of users were formed, each one with five test users, as described next:

Group 1: Students of software engineering, inexperienced in information retrieval systems, without a specific need of information.

Group 2: Authentic engineers of *ADD Intelligence in Aviation*, with specific needs for information to do assigned inspections.

Tests where organized in two phases. Phase one corresponds to a short query; phase two corresponds to long queries. This organization allows evaluation of relevant documents obtained by *Be Intelligent* system. Each phase consisted of the next steps: a) access *Discover* option in *Be Intelligent* system (a system screenshot is shown in Fig. 2); b) search for a concept or phrase depending on phase a system screenshot is shown in Fig. 2); c) review retrieved documents; d) record count of total documents found, count of total relevant documents identified, and position of most relevant document retrieved in a web form after each query.

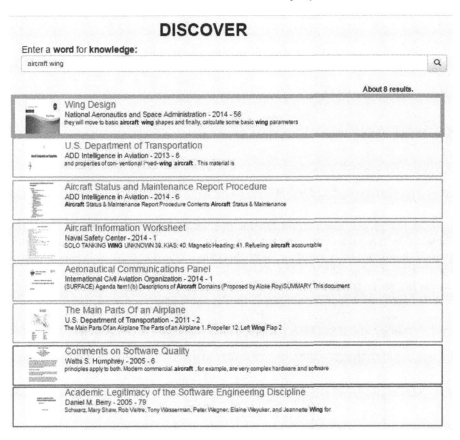

Fig. 2 Screenshot showing documents retrieved by Be Intelligent system when "aircraft wings" terms are searched

5 Results

Results obtained for each group and each phase (described in Section 4), are shown next.

5.1 Group 1 results

Phase 1. Fig. 3. shows calculated *Precision* to Group 1-Phase 1. No search obtained *1.0*, and average *Precision* was *0.73*. Best results were obtained by test users three and five, with a value of *0.78*. This value is considered respectable because the query domain of this group was wider.

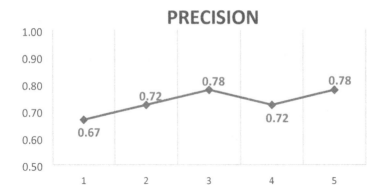

Fig. 3 Calculated Precision to Group 1, phase 1

Mean Reciprocal Rank calculated for queries of Group 1-Phase 1 is shown in **Table 1**. In this table is noted that first relevant document was located in third position, in queries three and four. With this data, *MRR* was *0.27*. This value was far from the ideal value of *1.0*, but it should be considered that this group experience level is low so its queries were less specific and the query domain was wider. Due to this, *Be Intelligent* responded adequately in this context.

Table 1. Calculated Mean Reciprocal Rank with results of Group 1-Phase 1

User	Rank	RR
1	4	$1/4 = 0.25$
2	5	$1/5 = 0.20$
3	3	$1/3 = 0.33$
4	3	$1/3 = 0.33$
5	4	$1/4 = 0.25$
	MRR	$= 0.27$

Phase 2. Results obtained with a long query done by Group 1-Phase 2, are shown in Fig. 4. In this figure can be observed that users one and two got *1.0* precision, while the average precision value is *0.89*, it is very close to the ideal value of *1.0*.

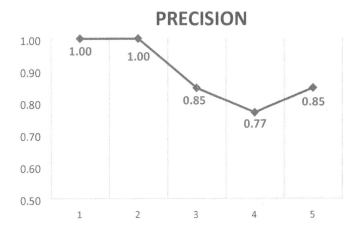

Fig. 4 Calculated Precision to Group 1-Phase 2

MRR for retrieved document in this phase was *0.29*, as shown in **Table 2**. Results with best-positioned documents were queries one, three, and five, where the first document was found in third position. *MRR* value is expected because the system does not respond to long queries with a group of a low level of experience.

Table 2. Calculated Mean Reciprocal Rank with results of Group 1-Phase 2

User	Rank	RR
1	3	$1/3 = 0.33$
2	5	$1/5 = 0.20$
3	3	$1/3 = 0.33$
4	4	$1/4 = 0.25$
5	3	$1/3 = 0.33$
	MRR	$= 0.29$

5.2 Group 2 results

In this section, are shown results obtained from experiment with Group 2. Authentic engineers formed this group, and due to this, their level of experience with search systems is high, as well as their clear information needs.

Phase 1. In this phase, the established query was directive FAA-2013-0695, a very specific term. Average Precision was *0.67*, and users three and four obtained *1.0* Precision value. However, users two and five obtained precision values under average because they needed more details about the directive, as shown in Fig. 5. Accordingly with this figure, a 67% of retrieved documents were relevant, and due to this, the system responded adequately during this part of an experiment.

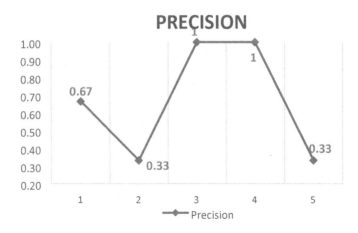

Fig. 5 Calculated precision to Group 2, phase 1

From data presented in **Table 3**, it is shown that *MRR* is *1.0*, because all test users indicated that first document shown was the most relevant for this query. The first retrieved document was judged as useful; for this reason, the system is judged effective.

Table 3. Calculated Mean Reciprocal Rank with results of Group 2-Phase 1

User	Rank	RR
1	1	$1/1 = 1$
2	1	$1/1 = 1$
3	1	$1/1 = 1$
4	1	$1/1 = 1$
5	1	$1/1 = 1$
	MRR	$= 1.0$

Phase 2. Data are shown in Fig. 6, the average precision for this phase of the experiment was *0.83*, indicating that more than 80% of retrieved documents are relevant, improving engineer decision-making.

PRECISION

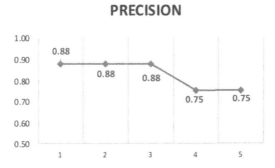

Fig. 6 Calculated precision to Group 2-Phase 2

In **Table 4**, are shown *RR* and *MRR* calculations. As can be observed, in 80% of queries document with most relevant information for the query is located in the first position. Due to this, the engineer found relevant information without loosing too much time reviewing other non-relevant documents.

Table 4. Calculated Mean Reciprocal Rank with results of Group 2-Phase 2

User	Rank	RR
1	1	$1/1 = 1$
2	1	$1/1 = 1$
3	2	$1/2 = 0.5$
4	1	$1/1 = 1$
5	2	$1/2 = 0.5$
	MRR	$= 0.80$

6 Conclusions

Computer systems to retrieve information from a large collection of documents must be able to reduce search time and allow classification of recovered information at the lowest cost, but with high quality, precision and validity to be useful to decision-making. *ADD Intelligence in Aviation* needed such a system, and in this research, results from the implementation of an inverted index data structure in the *Be Intelligent* system were shown. Two main contributions can be identified for *Be Intelligent* system: first, the creation of a storage and indexing scheme that facilitates information management of documents of routine use in a company. The indexing scheme supports workflow to include new documents in the indexing structure, the core of *Be Intelligent* system.

Second, an easy, quick and precise method to access their document collection was provided. Users now can retrieve information needed to achieve an aircraft inspection, without the extra steps required to do a search on an external hard drive and without loosing too much time reviewing non-relevant documents.

In this way, the application of an inverted index is helping a real company to achieve its duties in a more productive form.

We can describe one future direction to update the project, improvements of retrieval information with an algorithm for a passage retrieval approach. It could consist of four points: 1.- Execute a full document retrieval, 2.- Split the top retrieved documents into passages, 3.- Execute passage retrieval against the passage set created at point 2 and 4.- Return the top retrieved passages. To decide whether a passage retrieval strategy is useful or not, it is necessary to evaluate their ability to mine passages effectively.

References

[1]   C. D. Manning, P. Raghavan, and H. Schütze, *Introduction to Information Retrieval*. Cambridge University Press, 2008.

[2]   J. A. Arévalo, "Gestión de la Información, Gestión de Contenidos y Conocimiento," in *II Jornadas de Trabajo del Grupo SIOU*, 2007, pp. 1–15.

[3]   P. Lara Navarra and J. A. Martínez Usero, *Agentes Inteligentes en la Búsqueda y Recuperación de Información*, Segunda Ed. Barcelona, España: Planeta-UOC, S. L., 2006.

[4]   D. C. Blair, "The data-document distinction revisited," *ACM SIGMIS Database*, vol. 37, no. 1, pp. 77–96, 2006.

[5]   R. Arquero Avilés and J. A. Salvador Oliván, "La Investigación en Recuperación de Información: Revisión de Tendencias Actuales y Críticas," *Cuad. Doc. Multimed.*, no. 15, pp. 2–3, 2004.

[6]   F. J. Martínez Méndez, *Recuperación de información: Modelos, Sistemas y Evaluación*. Murcia, España: KIOSKO JMC, 2004.

[7]   M. Levene, *An Introduction to Search Engines and Web Navigation*. Wiley Publishing, Inc., 2010.

[8]   J. D. Anderson, "Guidelines for Indexes and Related Information Retrieval Devices," Bethesda, MD, 1997.

[9]   C. Martín-Daucasa, "Desing and Evaluation of new XML Retrieval Methods and their Application to Parliamentary Documents," Universidad de Granada, 2012.

Mobile Application for Automatic Counting of Bacterial Colonies

Erika Sánchez-Femat[1]., Roberto Cruz-Leija[1], Mayra Torres-Hernández[1] and Elsa Herrera-Mayorga[1].

[1]Unidad Profesional Interdisciplinaria de Ingeniería Campus Zacatecas – IPN, Blvd. del Bote S/N Cerro del Gato Ejido La Escondida, Col. Ciudad Administrativa, Zacatecas, Zacatecas, 98160, México
{erikasafe,robanac,matorres.26}@gmail.com,
veronica_qfb@hotmail.com

Abstract. In the following article it is proposed the design and implementation of a mobile application using a computer vision system that allows to count bacterial colonies in microbial cultures, decreasing significantly the time of quantification and generating a standard counting method for mobile devices running Android OS.

Keywords: Android, mobile application, microbiological count, computer vision

1 Introduction

The microbiological area forms part of the daily activities, everyday the microorganisms are participating in a beneficial and detrimental way, developing different roles. In the biological area, one of the tasks is the isolation and identification. Depending on the type of sample and analysis, it is necessary to know the number of colony forming units (CFU) present, this number meets the standards established under regulations in the area of food, health, water analysis, air and soil, among others. For these analyzes, experts of these areas perform manual counts that consume large amounts of time, yielding results that vary according to who performs the counting.

In this work it is proposed a mobile application to automate the quantification of bacterial colonies on culture plates, using a computer vision system, in which the image is cropped, converted to grayscale, contrast is improved and the image is segmented for counting.

This article is organized as follows: The section 2 describes the traditional counting method used in the microbiology laboratories, as well as some tools to facilitate the task; the section 3 shows the structure of the mobile application, explaining the stages of data acquisition, preprocessing and counting; the section 4 explains the results obtained using simulated and experimental samples of bacterial colonies; and finally the section 5 develops the conclusions of the work.

J. Mejia et al. (eds.), *Trends and Applications in Software Engineering*, Advances in Intelligent Systems and Computing 537, DOI 10.1007/978-3-319-48523-2_21

2 Traditional Counting Method

It is not only important to know the potential microorganism to cause a severe infection or a beneficial effect, but it is also important to know the number of microorganisms involved [1].
There are different types of methods or techniques to quantify microorganisms, which are described below:

i. Plate count method: This method has the advantage of having a good detection limit, however, it consumes larges amounts of time during the preparation of the plates [2].

The Mexican official standard NOM-092-SSA1-1994, Goods and Services. Method for aerobic bacteria count in plate [6] suggests using the following tools for counting bacterial colonies:

- Dark field colony counter, with adequate light, grid plate glass, and magnifying lens: It allows counting colonies per pulse of the counter, recording the result on a digital display.

- Mechanical or electronic recorder: It combines the function of electronic counter of pressure, with permanent marker labeling, preventing double counting.

ii. Petrifilm system: Its design has a rehydratable film coated with nutrients and gelling agents. Provides results in three steps: inoculation, incubation and counting [3].

3 Structure of the Proposed Application

The application can run in mobile devices with the following minimal characteristics:

- Android OS 4.1+
- 1Gb RAM
- 1.2GHz processor

3.1   General Operating Diagram

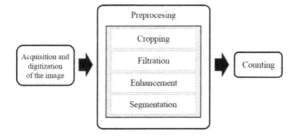

Fig. 1. General sequential operation scheme

3.2   Image Acquisition

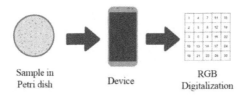

Fig. 2. Schematic operation of the data acquisition stage

In the figure 2 is shown the diagram of the image acquisition process of the application, in which is shown how the device takes the image and digitalizes it to RGB for future preprocessing.

3.3   Preprocessing

This stage involves the orderly implementation of filtering algorithms and improvement of existing images, where each has a well-defined objective.

Description of the preprocessing of the image.

The image conditioning is performed in four steps: cropping, filtering, enhancement and segmentation. Figure 3 shows graphically the order in which the process is performed, later it is explained how each of the stages are performed:

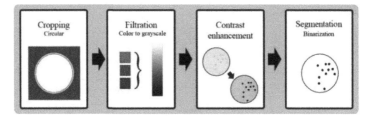

Fig. 3. Diagram of the preprocessing stages.

- Cropping: In the figure 4 is shown an example of an image processed by the mobile application, it can be seen that bacteria are inside a Petri dish, behind the plate can be seen a background which can change color depending of the environment where the photography is taken.

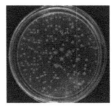

Fig. 4. Image capture by the mobile device

The bacterial colonies to count are concentred only on the content of the Petri dish, therefore, the areas relating to the outside are not useful and have to be excluded to prevent provide information of no interest in the stage of counting. An example of result of the stage of cropping is shown in the figure 5.

Fig. 5. Resulting image of the cropping stage

- Filtration: Within the capture and digitalization of images, samples are stored in RGB color model [5], the information stored in the different color channels Red, Green and Blue is not necessary to use it fully as it can be translated into excessive processing. In this stage of filtering a scale change is proposed in each pixel values, convert an RGB representation to grayscale. All this help to reduce the amount of information to process and facilities the stage of segmentation. An example in changing levels of color to grayscale is shown in the figure 6.

Fig. 6. Original image (left) in RGB format and resulting grayscale image (right)

- Enhancement: This stage is focused on improving the contrast in order to increase the difference between the bacterial colonies and the bottom of the Petri dish, we would expect the generation of a good segmentation stage following the application of contrast enhancement. The technique applied was linear expansion of the histogram [8], this process is based on a transformation of the gray levels, a linear distribution of the values that are within the range of 0 to 255 is performed.

In the figure 7 is shown an image without the application of the linear expansion and its corresponding histogram of gray levels, later in the figure 8 is shown the result of the implementation of linear expansion (contrast enhancement) and the resulting histogram of gray levels.

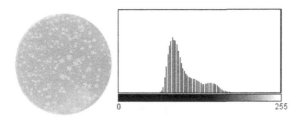

Fig. 7. Original image (left) and its histogram of gray color levels (right)

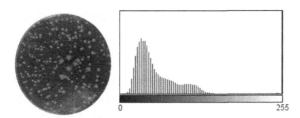

Fig. 8. Resulting image after applying linear expansion of the histogram (left) and its histogram of gray color levels (right)

- Segmentation: Before the counting process it is necessary to fully highlight the bacterial colonies with respect to the bottom of the Petri dish, for this, the Noboyuki Otsu's automatic binarization technique [4] is applied to the resulting image of the enhancement stage, the main goal is to generate sufficient discrimination to detect and segment the colonies. In the figure 9 can be seen the result of the implementation of the automatic binarization.

Fig. 9. Result obtained from implementing automatic binarization

3.4   Counting

To count the number of bacterial colonies in the image the Euler's method is used [7], where, the segmented image, is inverted and tagged. In the process of inverting the image, the colonies that were black are transformed to white. Then in the phase of tagging, tags are assigned to all objects found in the inverted image. These new objects will be now the bacterial colonies. The background of the image will not be tagged as a colony because it will have a different color, in this case white. Therefore, the number of colonies will be equal to the number of tags obtained. In the figure 10 can be seen the inverted image.

Fig. 10. Result of the process of invert the image for counting

4   Results

Then the results of the colony counts are presented from experimental and simulated tests achieved with this application called CA-BACT and the following two counting applications selected in the market (Table 1):

Name	Type of Application	O.S.	Counting area
CFU Scope[1] version 1.0.0	Mobile	iOS	Microbiology
AstroimageJ [2] version 2.4.1	Desktop	Windows	Astronomy
CA-BACT version 1.0	Mobile	Android	Microbiology

Table. 1. Comparison of selected applications

4.1   Simulated samples

In the first test, ten images were generated using the simulator (Figure 11) developed to test the application proposed using a sample of *Escherichia Coli* in agar culture medium with eosin methylene blue (EMB):

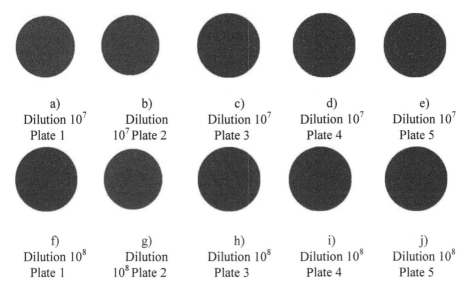

a)
Dilution 10^7
Plate 1

b)
Dilution
10^7 Plate 2

c)
Dilution 10^7
Plate 3

d)
Dilution 10^7
Plate 4

e)
Dilution 10^7
Plate 5

f)
Dilution 10^8
Plate 1

g)
Dilution
10^8 Plate 2

h)
Dilution 10^8
Plate 3

i)
Dilution 10^8
Plate 4

j)
Dilution 10^8
Plate 5

Fig. 11. Simulated samples

The results achieved by making automatic counts with the three applications were as follows:

---

[1] http://medixgraph.com/cfuscope
[2] http://www.astro.louisville.edu/software/astroimagej/

Image	Number of CFU simulated	CFU count with CA-BACT	CFU count with CFU Scope	CFU count with AstroImageJ
1)	81	80	31	81
2)	43	42	17	42
3)	53	52	13	54
4)	64	64	16	64
5)	70	69	21	71
6)	28	28	5	27
7)	36	34	11	36
8)	25	25	8	24
9)	30	30	10	30
10)	38	38	22	39

Table 2. Results of different automatic counts from simulated images

In the Table 2 can be seen the counting results of the three applications of which the count efficiency percentage was obtained as follows:

$$\text{Efficiency Percentage} = \frac{\sum_{1..n} CFU\ loss\ percentage}{n}$$

Where:

n is the number of selected samples

$$\text{CFU Loss Percentage} = \frac{(Simulated\ CFU - CFU\ automatic\ count) * 100}{Simulated\ CFU}$$

Given that the application CA-BACT has a counting effectiveness rate of 98%, the application CFU Scope has an effectiveness of 32% and the AstroimageJ application has a 100% of effectiveness.

Allowing establish that although the application AstroimageJ obtained a better result with the simulated images, the proposed application CA-BACT has the advantage of being a mobile application which potentiates its usability features, an easy download and installation from the Android Application Store (Google Play Store), more promotion and outreach to potential users.

For all the above it is concluded that the image preprocessing and the counting algorithm of the proposed application is efficient and is developed in a technology that gives preferred use and more visibility to the user.

4.2   Experimental samples

The samples used in these experimental tests were generated in the laboratory of Biology of the Interdisciplinary Professional Unit of Engineering Campus Zacatecas (UPIIZ), having four test samples of effluent dams that contained *Escherichia Coli* in differential culture medium EMB (figure 12), the photographs were taken with a standard device[3] in uncontrolled lighting, background, perspective and reflection conditions, having the following results with reference to the average manual count by four users of the same laboratory, the Table 3 shows the results:

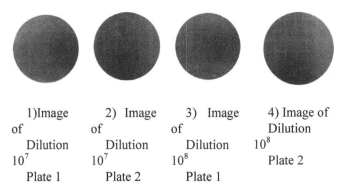

| 1)Image of Dilution 10^7 Plate 1 | 2) Image of Dilution 10^7 Plate 2 | 3) Image of Dilution 10^8 Plate 1 | 4) Image of Dilution 10^8 Plate 2 |

Fig. 12. Experimental samples

#	Avg. CFU manual count	Avg. manual count time	Counting with CA-BACT	Counting with CFU Scope	Counting with AstroImageJ
1)	57	3 min	73	18	423
2)	91	4 min	75	43	567
3)	35	2min	22	12	398
4)	27	1 ½ min	21	9	400

Table 1. Manual counting results against automatic counts in the experimental test.

In this test for the counting efficiency of the three applications, the average CFU number of the four manual counts is taken as parameter to compare, having the following results: The proposed application CA-BACT has a counting effectiveness rate of 74%, the application CFU Scope has an effectiveness of 32% and AstroimageJ has a -394% of effectiveness, this negative value shows that the application AstroimageJ it is not feasible with noisy images, because it does not perform any preprocessing to the image.

In the Table 3 can be seen that the times when manual counting is performed in valid samples of 25 to 250 CFU can vary from one minute and a half to about five minutes per sample, having a major error in the manual counting of samples with more number of CFU, so an automatic counting saves almost 98% of time as automatic counting applications presented here take between 2 and 5 seconds to count, and also completely eliminate the difference of counts in the same sample, because

---

[3] Device with 1GB RAM, 1.2GHz Dual processor and 8Mpx camera

during the experimental tests it was observed that in the manual count of a sample more than once with the same person can produce different results, instead, when the application uses the same image, it always gives the same result.

5   Conclusions

CFU losses in the count with the developed application CA-BACT can be lessened by controlling the shooting conditions of the image that is processed by the application such as lighting, background, perspective and reflections.

In the area of computer vision other techniques can be used for transforming the intensity of the images so it can give better results with the proposed filters in this paper.

A good image preprocessing serves a very important role within the application, as it helps to have better results than if the image was processed as obtained in the acquisition stage.

The proposed mobile application was tested with volunteer users of the biology laboratory at UPIIZ to check its efficiency and effectiveness. It's about to start the implementation of the general use with all users of this laboratory; however, by following norms and standards used in this type of laboratories such as The Mexican official standard NOM-092-SSA1-1994, when counting in culture plates, this mobile application can be a general purpose application in this area.

6   References

1. Corral-Lugo, A, Morales-García, Y, Ramírez-Valverde, A, Martínez-Contreras, R and Muñoz-Rojas, J.: Cuantificación de bacterias cultivables mediante el método de "Goteo en Placa por Sellado (o estampado) Masivo". Revista Colombiana de Biotecnología vol., no. 2 (2012)
2. Ortega Olguín, I.: Comparación de métodos de cuantificación de bacterias lácticas expuestas a estrés y durante su desarrollo en salchichas (2014)
3. Alonso Nore, L., Poveda Sanchez, J.: Estudio comparativo en técnicas de recuento rápido en el mercado y placas Petrifilm 3M para el análisis de alimentos. Universidad de Bogotá (2008)
4. Nobuyuki, O.: A Treshold Selection Method from Gray-Level Histograms. IEEE Transactions On Systems, Man, And Cybernetics, pp.62-68 Vol. SMC-9, No. 1 (1979)
5. L. Saphiro, G. Stockman.: Computer Vision. Prentice Hall. pp. 213-215 (Jan 1, 2001)
6. Norma Oficial Mexicana NOM-092-SSA1-1994.: Bienes y servicios. Método para la cuenta de bacterias aerobias en placa. Diario Oficial de la Federación (1995)
7. Fuente López, E, Trespaderne, F.: Visión artificial industrial. Secretariado de Publicaciones e Intercambio Editorial (2012)
8. L. Deligiannidis, H. R. Arabnia.: Emerging Trends in Image Processing, Computer Vision and Pattern Recognition, Morga Kaufmann, Elsevier. Pag. 189 (2015)

Part IV
Information and Communication Technologies

A New Scheme to Visualize Clusters Model in Data Mining

Wilson Castillo-Rojas[,1], Fernando Medina Quispe[1], Juan Vega Damke[1]

[1] Arturo Prat University, Faculty of Engineering and Architecture,
Av. Arturo Prat 2120 Iquique, Chile
{wilson.castillo, jfranciscov7, femedina}@gmail.com

Abstract. This paper presents the design and implementation of a new scheme to visualize clusters model called IVCM, in the context of a data-mining process. The visualization of a cluster model becomes complex when the dataset is high volume, density and dimensionality. The design of IVCM scheme is based on four characteristics: interactive visualization, combination of models, ad-hoc graphics artifacts, and use metrics. The objective of this scheme is to contribute to the analysis and understanding of a clustering model. Metrics considered in this proposed scheme, allow comparison instances of different clusters, which in turn helps to understand how groups are composed. Through the implementation of a web visual environment that meets the characteristics defined in IVCM, and an online assessment of 23 users, positive results on the usefulness of this new visualization scheme are achieved.

Keywords: Cluster Visualization, model visualization for data mining, data mining visualization, interactive visualization of models, data visualization schemes.

1 Introduction

Understand and properly interpret the patterns resulting from a model, it is most critical in a Data-Mining (DM) process. It's argued that one of the ways to support this is by using appropriate visualization in model building. This establishes that knowing the inner workings of the model, allows on the one hand, understanding how it works and on the other hand, better interpret the results. In particular, visualization of cluster models is difficult to carry out, especially when the dataset is large volume, and the number of dimensions is high. For example, with 3 dimensions can't be observed all attributes simultaneously, this it's more complex in the case of high-density clusters.

The review of the state of art in this work, and evaluation of current DM tools, it was found the difficulties presented by visualizing the clustering models. Also, there are various metrics to compare clusters, but most of the tools use the distance calculation, and very few, implement comparison instances. In addition, DM tools

---

[1] Please note that the LNCS Editorial assumes that all authors have used the western naming convention, with given names preceding surnames. This determines the structure of the names in the running heads and the author index.

© Springer International Publishing AG 2017
J. Mejia et al. (eds.), *Trends and Applications in Software Engineering*, Advances in Intelligent Systems and Computing 537, DOI 10.1007/978-3-319-48523-2_22

analyzed don't perform combination of techniques, and don't provide an appropriate level of interaction.

This research precisely addresses the problem of complexity for visualizing clustering models. For this, it's considered some needed elements for the design and development of a new visualization scheme, that allows to explore the model on demand, for which is fundamental to have appropriate interaction mechanisms. Another element is the combination of cluster model with a descriptive model to establish relationships between the attributes of the dataset. Also, it is necessary to have ad hoc graphical artifacts to visually represent clusters, particularly in dataset with high volume, density and dimensionality. Finally, to compare quantitatively instances of different clusters, and the cluster homogeneity with appropriate metric for this.

This article describes a new scheme to visualize clusters model, mainly when dealing with multidimensional data. Conceptual design and technical aspects are presented in addition to the implementation of a web visual environment, which provides the main features of this scheme called IVCM: interactive visualization to explore the model, combination of models to describe relationships between attributes, graphics for the clustering model, metrics to compare instances of different clusters and measure the level of compactness or dispersion of the clusters. Finally, the results of a subjective evaluation obtained through an online survey with 23 users with experience in DM processes are discussed.

2 Visualization in Data Mining

The DM process seeks to obtain, from a massive dataset, data models to describe or predict new instances. This involves steps of data preparation (input), processing of data to achieve models, and interpretation of the resulting patterns (output). This output should mean new knowledge in the organization, useful and understandable for end users, and can be integrated into processes to support decision-making. However, the difficulty is to identify patterns in the data, which is a complex task and often requires experience, not just the data analyst, but also the expert in the problem domain.

One way to support data analysis, models and patterns, is through its visual representation. Using the capabilities of human visual perception, which can detect patterns more easily. Under this approach, DM visualization has been mostly used in the descriptive analysis of the data (input) and presentation patterns (output), leaving limited this paradigm for modeling analysis [1].

Meneses and Grinstein discussed in [2], elements to move towards a schema for the DM process, with support for display four types of entities: data, algorithms parameters, induced models, and validation for patterns.

A key factor in DM, to improve predictive or descriptive ability of a model is to understand how the induced model is working. Among other aspects, for example; it is important to understand how the model performs a distribution of the dataset according to different attributes, how the instances of the model are correlated with a subset of instances, and how attribute values are partitioned by the model. Some machine learning techniques operate as closed systems, and it's difficult to achieve an interpretation of the patterns obtained (e.g., the weights of a trained artificial neural network), and answer user questions concerning processing performed by the model. A notable example is the artificial neural networks, which converge to a set of numerical weights that have no direct interpretation in the problem domain.

Techniques such as decision trees are easily understandable when the model is small, larger trees (over three levels) interpretation is complex. The same is true for cluster model, because with large data volume and high dimensionality, is not simple display all attributes simultaneously. Other learning techniques such as association rules, the size of the model also presents problems (long lists of rules), and requires the development of new graphics, providing better visualization and interaction with the model, for easy interpretation. It is considered that integration of visualization in the DM process can be of significant importance, as it can help in two ways:

• Provide visual understanding of complex computational approximations and;
• Discover complex relations between the data that aren't detectable with automated methods of analysis, but they can be captured by the human visual system.

From the foregoing, it follows that alternating iteration between visualization and automatic DM provides the data analyst, support in the task of pattern recognition. Including humans in this process, it achieved a good combination of visual perception, with the computing power and computer storage [2]. Today, there are few tools that incorporate model Visualization [3, 4]. Those that exist are very limited in their ability to explore, and answer questions about the changes, which performs the data model to generate patterns.

3 Visualization to Clusters Model

The cluster analysis is a task of DM which aims to identify and describe groups in a dataset, so that the elements assigned to the same group are similar to each other, while the elements belonging to different groups are dissimilar to each other. In general, the distance measure is most often used to make comparisons between the attributes of the instances. This task is used for lifting profiles: customers, employees, hospital patients, etc. At the end of the clustering process it is achieved as a result label each instance of the dataset, associating it with one of the generated groups [5].

There are several methods or techniques to generate clusters, and each uses a different principle of induction. Fraley in [6], proposes two types of clustering methods: hierarchical and partitional. Furthermore, Han and Kamber proposed [7], three types of clustering based on density, patterns and grids. There are various visualizations types to represent clusters models, among which:

• Scatterplot: is the most used graphic, and shows the relationship between two numeric variables. Each point corresponds to the values in the coordinates *(x, y)*. Each group is assigned a color and instances belonging to that group take that color, in this way it is easy to identify which group they belong. Instances of the dataset are plotted with up to three dimensions. Through this graph, we can identify correlations between variables.
• Voronoi diagram: it applies to techniques that generate centroids, and consists of a partition of Euclidean space. For each pair of centroids, perpendicular bisectors are created, which generate parallel segments between them. This type of cluster

visualization generates polygons for each group, where the instances are distributed inside.

- Groups density: corresponds to a visualization where you can see the densities of each group. Shows the region covering each cluster, and allows analysis of instances.
- Scatterplot matrix: is a generalization of a scatterplot in n dimensions. It allows to observe simultaneously the behavior of numeric variables in several dimensions. It's a scan tool data for comparing a set of instances with respect to all its attributes. This graph shows all combinations of scatterplot between two variables in a single view under a matrix structure. For n dimensions: n rows and columns are displayed.
- Parallel Coordinates: this type of diagram consists of a horizontal axis and several vertical axes. Each vertical axis represents a variable whose values will be represented along the horizontal axis. In the case of numerical variables, the values are sorted from lowest to highest (bottom-up). Each polygonal line in this diagram corresponds to an instance (row) of the dataset, and moves between the vertical axes depending on the values that corresponds to each of the variables. This chart, accompanied by interactive elements, provides a powerful tool for viewing multidimensional objects.
- RadViz: it's a visualization technique of radial coordinates. Shows all the attributes as anchor points circumference of a circle and spaced equidistantly depending on the number of attributes. Within the circle, the instances are shown as points, which are arranged on the graph. It builds on the paradigm of tensor from particle physics, points the same kind attract each other, the different kind repel each other, and the resultant forces exerted on the anchorage points. An advantage of RadViz is that it retains certain symmetries of established data, and its main disadvantage is overlapping points [8].
- Dendrogram: is the simplest way in which a hierarchical data structure can be represented, and corresponds to a tree-shaped graph (dendro, meaning tree). Data items are represented by the leaves in the final level of the tree structure, while nodes at the highest level, are representing the groups or clusters of data items with different levels of similarity (Jain and others [9]). The classic way to represent the dendrogram is draw as rooted tree with root anchored centrally in the top of the image, and the branches of the nodes-children down using straight lines or diagonal.
- Radar chart: Each instance is plotted as a polygon in a radial space, and its edges are generated in relation to the magnitude of its dimensions, instances belonging to same groups have the same colors. Geometrical figures presenting similarities in structure correspond to similar objects.

4 Related Works

Long proposed in [10], consistent work in a visualization solution for hierarchical clustering models. Its main hypothesis is that there are two major problems to visualize models clusters:

a) The large amount of data limit the display of a clustering model, as the screen space cannot observe all instances simultaneously.

b) When the number of dimensions is greater than 3, you cannot be displayed all attributes simultaneously.

The proposal consists of two steps: clustering and visualization. It helps data analysts to understand the distribution of a set of data with high dimensionality. In the first step, hierarchical algorithms techniques are used to generate clusters. In the second step, two methods are provided for displaying the results. The first method uses optimized graphic of stars, which minimizes the overlap. In the second method, visualization techniques are combined; parallel coordinates, radial for hierarchical structures, and parallel coordinates circular. In Figure 1, it can be observed using parallel coordinates linked to radial display hierarchical trees. The diagram of parallel coordinates located on the right side, is displayed interactively selecting clusters of the left side.

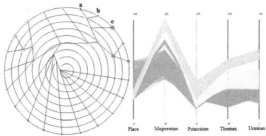

Fig.   1.   Parallel   coordinates
linked with radial visualization.

Lee and others proposed in [11], an interactive visual environment for clustering models, in a context of analysis of text documents, that generates groups based on the most repeated words in the set.

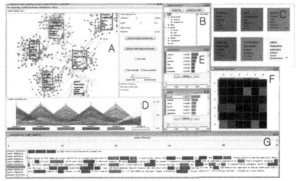

Fig.   2.   Visual
environment proposed by Lee and Others [11].

It provides various mechanisms of interaction to refine the results of a clustering process also allows data filtering out of range, and re-grouping of data. Figure 2 shows the main interface, in which there are various elements for cluster analysis: parallel coordinates, views of relationship for clusters, views of trees, among others.

As regards the use of metrics, for comparing cluster instances, Grabusts in [12] presents a job which performs an analysis of the results obtained by the K-means algorithm with a given dataset. For the generation of clusters, uses metrics: Euclidean distance, Manhattan distance and Pearson correlation coefficient. It's concluded in this study that the results are very similar to each metric used, and that the Pearson correlation delivers best values for all clusters.

Once realized the process of clustering, it's necessary to evaluate the results obtained to validate the quality of the model. Cluster analysis is considered as a process of unsupervised learning, i.e. the result cannot be compared with a value previously known. Thus the confusion matrices and precision metrics used in predictive tasks, do not apply to these cases. However, a method for evaluating the clustering process is through evaluation criteria, which are divided into internal and external. In the literature there are several criteria for evaluating clusters, however, only some of them are applied. These criteria include: the sum of squared error (SSE), the rate of Dunn, and Davies-Bouldin index [5].

On the other hand, five DM tools were reviewed in relation to clustering models, particularly: the display level, if it provides metrics, and the level of interaction in three types of selected graphics. The results are shown in Table 1.

The visualization level is qualified, considering the amount of visual elements that the tool has. If less than 2, your score is low, if equals 2 is scored regularly, and greater than 2 is good. And to measure the level of interaction in the graphs, one considers the number of interactions available, for each visual element: less than 3 is low, between 3 and 4 regulate, and greater than 4 is high.

Table 1. Features tools analyzed.

Tools / Features	Visualization level	Interaction mechanisms	Metrics
Orange	Regular	• Parallel coordinates: **Regular** • Radar graph: **N/A** • Scatterplot: **Regular**	Non
Weka	Low	• Parallel coordinates: **N/A** • Radar graph: **N/A** • Scatterplot: **Regular**	Yes
R	High	• Parallel coordinates: **High** • Radar graph: **Low** • Scatterplot: **High**	Yes
Knime	High	• Parallel coordinates: **Regular** • Radar graph: **Low** • Scatterplot: **Regular**	Non
SPSS Modeler	Regular	• Parallel coordinates: **Low** • Radar graph: **N/A** • Scatterplot: **Low**	Non

It can be seen from Table 1, R is the most complete tool, but doesn't have radar chart, and along with Weka are the only tools that provide metrics for cluster analysis. However, R is a programming environment that manages the generation of models and visualization, through code with the library called Ggobi [13]. This means that it's a DM tool, which is not easy to use. With regard to the proposed design of this research, it is mainly based on two approaches:

i) The field of Visual Analytics described by Keim et al [14], focuses on the process of analyzing and managing large volumes of data through integration of human judgment on visual representations and interaction mechanisms. It combines areas: visualization, DM, and statistics. In this approach, visualization is a central step in the process, which is oriented not only to the visual description of data (inputs), it's essential to build the model (process), and to represent the knowledge gained through patterns (departure).

ii) The second approach taken as a reference, corresponds to an augmented visualization scheme for models (VAM) in the DM process, proposed by Castillo and others [1]. Consider this scheme as central: visualization models for exploratory analysis, models of visual perception, interaction mechanisms, combining DM techniques, and metrics to compare model instances. Presents a model of visual perception and user interaction, focusing on the step of adjusting the model in a DM process, and establishes a way to explore the original model and each of its instances, considering the characteristics or recently indicated axes.

5 IVCM Scheme

The objective of this job, is to design and develop a scheme of Interactive Visualization for Clustering Models (called IVCM) for DM process. This scheme should enable the analyst: visually explore a model in its review phase. It's mainly aimed at dealing with dataset of high volume and dimensionality. It's based on two approaches: visual analytics [14] and augmented model visualization [1]. The IVCM scheme consists of two states: data visualization, and visualization of the clustering model, as shown in Figure 3.

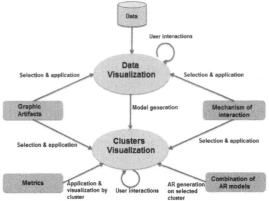

Fig. 3. IVCM Scheme.

- State 1 (data visualization): allows the data analyst: select, load and present the set of input data or minable view. The user can perform exploratory data analysis, using a set of graphical artifacts and appropriate interaction mechanisms. The dataset must be previously prepared, as this scheme IVCM doesn't consider the preliminary preparation of the minable view.
- State 2 (cluster visualization): the transition to the second state occurs when the data analyst requires the generation of the model. For this, the user must specify

the number of clusters as a parameter. Besides graphic devices, and interaction mechanisms, the user can use metrics to compare instances of different clusters, and also to measure the compactness or dispersion of each cluster. Additionally, the data analyst can generate a model of Association Rules (AR) on a selected cluster, in order to achieve a description that can determine relationships between attributes of the instances, and with this, you can better understand the conformation clusters.

For this, considered in its design the following characteristics:

- Graphical artifacts: it's necessary to have a set of graphical artifacts linked and suitable to represent clustering models. Mainly aimed at manipulating dataset with large volume and high dimensionality. As a result of the study of art and related work, are achieved determine the most appropriate visual elements to address the complexities outlined above, these are: matrix scatterplot, parallel coordinates, radar chart and circular.
- Interaction mechanisms: the analyst must be able to move through the clustering model to explore, as well as its instances. You should also be able to select instances, to compare them with each other. For this, it's necessary that the analyst can interact with graphics artifacts. Interaction mechanisms listed in Table 2 are considered.

Table 2. Interaction mechanisms in IVCM.

	Selection and highlight	Selection and link	Zoom and pan	Filter	Reorder	Detail
Parallel coordinates	Non	Non	Non	Yes	Yes	Non
Radar chart	Yes	Non	Non	Non	Non	Yes
Scatterplot	Yes	Yes	Yes	Yes	Non	Yes
Circular chart	Non	Non	Non	Non	Non	Non

- Combination of AR model: on the clustering model, it's possible to apply the technique RA to expand the description of the model, and determine relationships between attributes. The reason why has selected the RA technique is that it's a descriptive technique, which through rules allows additional insight of the formation of clusters. It's used for generating the clusters "K-Means" algorithm, and the "Apriori" algorithm to generate the RA.
- Metrics: the results of clustering processes are compared through internal assessment criteria, which consider the distances: intra-cluster and inter-cluster. This allows to quantitatively compare instances of clusters, and measuring the level of homogeneity of the clusters. With this, knowing the conformation of the groups generated by the process. From the study of metrics for clustering models, three measures are selected: the sum of the mean square error (SSE), Davies-Bouldin index (DB-Index), and the Euclidean distance. SSE y DB-Index are implemented to measure the level of compactness or dispersion of the clusters, and the third metric is used to generate clusters.

6 Visual Environment

The visual environment implements the scheme IVCM under a client-server web architecture. The client side is responsible for generating visualizations and interactions, and is on the server where clustering and AR algorithms are executed (K-Means, Apriori), and metrics are also calculated. The visual environment is primarily developed using JavaScript both on the client and server. Its graphic and interaction engine is D3.js. To calculate the metrics and make calls to DM algorithms, uses Node.js.

In Figure 4, the main interface visual environment once the user has selected and loaded the dataset is presented to analyze. This part represents the first state IVCM schema (data visualization). In this state, the analyst can perform multidimensional analysis of data and can select multiple instances to be compared. The scatterplot and parallel coordinates are linked. Also, in these graphs, the user can perform interactions described in Table 2. The dispersion matrix is presented in a diagram, the user can select the attributes required for analysis, through the coordinates *(x, y)*, and linked graphics are updated simultaneously on the screen.

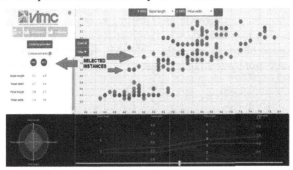

Fig. 4. Visualization Data Interface (state 1).

It can also be seen in this same figure, two selected instances, and which are indicated with green arrows, in the scatterplot. Also, they are highlighted simultaneously in the graph of parallel coordinates, which appears at the bottom of the screen. The values of the attributes of selected instances, are added to the comparison area on the left side of the interface, as is also indicated by a green arrow, and are displayed on the radar chart in the lower left corner. Thus, it's possible to quantitatively compare two or more instances, in order to find differences or similarities in their characteristics.

Once the user clicks the button to generate the clustering model, it is passed to the second state in the IVCM scheme (cluster visualization), and the interface is presented in Figure 5. It can be seen; a new section appears on the right side of the screen. This section shows, complementary views that are provided by: the description of the cluster next to the value of metrics (SSE and DB-Index), the radar chart, and AR generated on the clusters that the user chooses.

In this second state, in addition to graphics and interaction mechanisms, the analyst has a view of the multidimensional clustering model, through a matrix scatterplot that can be configured according to dimensions or attributes that requires analysis. As in the first state, this diagram is still linked to the parallel coordinates graph which is presented at the bottom of the screen.

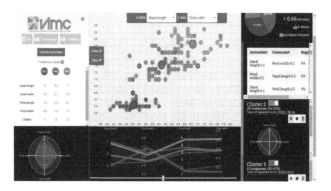

Fig. 5. Clustering visualization interface (state 2).

Additionally, the user can obtain additional views on the clustering model, with the application of a AR technique on each cluster, which allows to describe existing relationships between attributes of the instances. The RA technique is available for any dataset, and the user can provide as a parameter the percentage confidence and required coverage. You can also add a radar chart for each cluster and with this, you can complement the exploratory analysis of the AR, for example you can determine which attributes have more weight in a cluster.

7   Evaluation and Analysis of Results

7.1   Controlled Experiment

To estimate the practical value of the characteristics defined in the IVCM scheme, it has carried out an experiment to get a subjective perception of its usefulness. For this, we designed a DM task, which is to generate a clustering model from a given dataset.

This experiment was conducted with a universe of 23 people, which have different levels of experience about: DM processes, and the use of DM tools. All the participants have computer training level undergraduate and graduate students.

The experiment was carried out in 3 stages:

- First, users access the following URL: http://vimc.inf.unap.cl. In this direction, they find a tutorial with animation explaining the research, access to the visual environment, and also the online survey that should respond, once analyzed the clustering model presented.
- First Stage - Tutorial: at this stage, the research is explained, and the main features about the IVCM scheme. Also, the user knows the general aspects of the visual environment.
- Second Stage - Training: the user works with the visual environment and a set of training data, where it can interact freely generating a clustering model, compare instances, and apply AR in the model. The objective of this stage is to familiarize the user with the visual environment, and usability that provides for: generate, explore and analyze a clustering model.

• Third Stage - Evaluation: in this third stage is presented to the user, the result of the clustering with a second set of data, which were generated 2 clusters. It has also been applied to the first cluster technique RA, and two items have been selected (one from each cluster). This stage aims, the user must observe and analyze the clustering model presented, then answer an online survey, which evaluates the IVCM scheme and the visual environment.

The dataset used in steps 1 and 2 is known as Iris, and contains 150 records three kinds of flowers. Where the first 50 records are the Setosa type the following 50 registers are Virginica, and the last 50 records are Versicolor. Each record has 4 attributes (length of petal, sepal length, width petal, and sepal width). The dataset called Wine was used to stage 3. This dataset contains 1000 records of wines with 12 attributes or characteristics (fixed acidity, volatile acidity, citric acid, residual sugar, chlorides, free sulfur dioxide, total sulfur dioxide, density, pH, sulfate, and degree of alcohol). However, the visual environment can handle any data set, and is only required to be in the open CSV format.

The survey designed to validate the work consists of 22 questions divided into 6 sections: participant information, visualizations, analysis of model clustering, AR analysis, comparison of objects, and the IVCM scheme.

7.2 Evaluation of Results

All the participants in the experiment, express a high assessment of IVCM scheme and visual environment. Regarding the three features of the proposed scheme, most have evaluated the tool as very good with 70%, and good in 30%, relative to the level of interactions They provide its visual elements. The combination of DM models is well regarded by users, with 30% very good and 57% good, and the same valuation obtained metrics that compare the compactness and dispersion of clusters.

Regarding the utility level of visual elements available to describe both data and instances in clusters, that users deliver the highest valuation for the scatterplot, with 87 % positive appreciation (61% very good and 26% good), second, the radar chart, with 73% (43% divided into very good and 30% good). And third, parallel coordinates, with 61% positive rating (9% good and 52% very good), and also get most negative assessment (26% regularly and 13% bad).

When asked participants if the generation of a AR model on a cluster, helps improve understanding of these, 95% said yes, while only 4% said the opposite. When they asked; if the comparison of instances helps to understand better how the clusters are composed, 35% answered totally agree, 57% agree, and only 9% disagree. As for the metrics available, 30% of users think that their usefulness is very good, 57% think it's good, only 13% think it's fair, and no one thinks that is bad or very bad.

One of the tasks of the experiment aims to validate the visual analysis that gets the user with the tool. And for this, the user must observe two clusters, and then determine which of the two is more dispersed. His response is compared to the metric (SSE), which empirically determine the dispersion of the clusters. It's obtained as a result that 91% of participants, was able to respond correctly that cluster has greater dispersion, corresponding to the second cluster presented. This is corroborated with its value for SSE of 15.057, compared with the first cluster that has only one value for SSE of 10.806.

Finally, faced with the question: if the IVCM scheme supports the understanding of a clustering model; 48% of participants answered totally agree and 52% agree. This means that all participants, positively evaluated the proposed visual scheme.

8 Conclusions and Future Work

The conclusions that have been obtained in this work are the following:

In general, regarding the results of research work, these are very encouraging as most of the participants in the experiment, give a positive evaluation of the characteristics of IVCM scheme.

The visual environment achieves a positive evaluation regarding: usability, interaction mechanisms, and implemented graphics. The latter, allow appropriate exploratory analysis of the dataset, and also for a cluster model with multidimensional data. It highlights the high valuation of the interactions implemented in each visual element, and also between graphics devices, the scatterplot is the one that gets better evaluation.

Regarding the selected and implemented metrics, although their assessments are good, are not those expected, so it's considered an opportunity for improvement for next version of the visual environment, particularly as it relates to display these metrics.

Regarding combine DM techniques, although it's positively assessed the implementation of AR on each cluster, and serve to improve the description of the groups, the idea is to improve and assess their visual representation, so that be more understandable by users.

Finally, the IVCM scheme through its visual environment, achieves the goal of supporting the understanding of a clustering model. Visual analysis of user experimentation was corroborated with metrics, and both measures coincide, for example, recognize whether a cluster is more compact than other.

With regard to future work, this is related to the following aspects:

Evaluate other descriptive DM techniques that provide additional visualizations to the AR. Also, the to visualize generated models RA technique.

Selecting and applying new graphical artifacts that may be more useful in exploring data, and clusters, preferably for dataset with high dimensionality. An interesting improvement is to integrate all DM algorithms that provides Weka.

References

1. Castillo, W., Meneses, C., Medina, C.: Augmented visualization for data-mining models. Journal: Elsevier Procedia Computer Science. ISSN: 1877-0509, DOI: 10.1016/j.procs.2015.07.063, volumen 55, pp. 650-659 (2015).

2. Meneses, C., Grinstein, G.: Visualization for Enhancing the Data Mining Process. In Proceedings of the Data Mining and Knowledge Discovery: Theory, Tools, and Technology III Conference. Orlando, FL. (2001).
3. Keim, D., Kohlhammer, J., Geoffrey, E., Mansmann, F.: Mastering the Information Age Solving Problems with Visual Analytics. Edited by the authors Published by the Eurographics Association Postfach 8043, 38621 Goslar, Printed in Germany, Druckhaus Thomas Müntzer GmbH, Bad Langensalza. Theoretical Issues in Ergonomics Science. Vol. 8, No. 1, ISBN 978-3-905673-77-7, (2010).
4. Castillo, W., Meneses, C.: A Comparative Review of Schemes of Multidimensional Visualization for Data Mining Techniques. III Congreso Internacional de Computación e Informática del Norte de Chile (INFONOR-CHILE). Arica – Chile (2012).
5. Maimon, O., Rokach, L.: Data Mining and Knowledge Discovery Handbook, 2nd ed. Springer Science+Business Media, Inc. Edited by Oded Maimon and Lior Rokach Tel-Aviv University, Israel. ISBN 978-0-387-09822-7, (2010).
6. Fraley, C., Raftery, A.: Model-based methods of classification: Using the mclust Software in Chemometrics. University of Washington Seattle, United States. DOI: 10.18637/jss.v018.i06, (2007).
7. Han, J., Kamber, M.: Data Mining: Concepts and Techniques, 2a ed., Estados Unidos, Elsevier. ISBN-10: 1558609016. ISBN-13: 9781558609013, (2006).
8. Hoffman, P., Grinstein, G.: A Survey of Visualizations for High-Dimensional Data Mining, in: Fayyad U., Grinstein G. G., Wierse A. (eds.), Information Visualisation in Data Mining and Knowledge Discovery, Morgan Kaufmann Pub., San Francisco, pp. 47-85, (2002).
9. Jain, A., Murty, M., Flynn, P.: Data Clustering: A Review, ACM Computing Surveys, Vol. 31., No. 3., pp. 264-323, (1999).
10. Long, T.: Visualizing High-density Clusters in Multidimensional Data. Bremen, Germany. DOI 10.1007/s00180-011-0271-3, (2011).
11. Lee, H., Kihm, J., Choo, J., Stasko, J., Park, H.: iVisClustering: An interactive visual document clustering via topic modeling. In Computer Graphics Forum (Vol. 31, No. 3pt3, pp. 1155-1164). Blackwell Publishing Ltd., June (2012).
12. Grabusts, P.: The choice of metrics for clustering algorithms. Letonia. ISSN 1691-5402, ISBN 978-9984-44-071-2, 2011.
13. Dianne, D., Deborah, F.: Interactive and Dynamic Graphics for Data Analysis: With R and GGobi, ISBN 978-0-387-71762-3, 2007.
14. Keim, D., Kohlhammer, J., Geoffrey, E., Mansmann, F.: Mastering the Information Age Solving Problems with Visual Analytics. Edited by the authors Published by the Eurographics Association Postfach 8043, 38621 Goslar, Printed in Germany, Druckhaus Thomas Müntzer GmbH, Bad Langensalza. Theoretical Issues in Ergonomics Science. Vol. 8, No. 1, ISBN 978-3-905673-77-7, 2010.

By Clicking Submit Button, You will Lose Your Privacy and Control Over Your Personal Information

Rafael Martinez-Pelaez[1], Francisco R. Cortes-Martinez[2],
Angel D. Herrera-Candelaria[2], Yesica I. Saavedra-Benitez[3], and
Pablo Velarde-Alvarado[4]

[1] Facultad de Tecnologías de Información, Universidad de la Salle Bajío, León, México
rmartinezp@delasalle.edu.mx
[2] División Multidisciplinaria de Ciudad Universitaria, Universidad Autónoma de Ciudad
Juárez, Ciudad Juárez, México
{al114359, al114847}@alumnos.uacj.mx
[3] División de Estudios de Posgrado e Investigación, Instituto Tecnológico de Toluca, Toluca
de Lerdo, México
ysaavedrab@toluca.tecnm.mx
[4] Área de Ciencias Básicas e Ingenierías, Universidad Autónoma de Nayarit, Tepic, México
pvelarde@uan.edu.mx

Abstract. Since the beginning, people have perceived privacy and security as the biggest risk related with Internet. As a result, many Internet security solutions have been developed, such as antivirus, authentication mechanism, firewalls, security protocols, and security policies. However, every year organized crime and criminals commit several attacks against innocent people via the Internet. In this paper, we explain how millions of Internet users lost their privacy and control over their personal and financial information by means of the web registration form and social network sites.

Keywords: authentication, cyber-crime, fraud, identity theft, privacy, spoofing, web registration form.

1 Introduction

Today, the Internet is more and more important in our daily lives. This means that, we require internet connection to do many activities, such as communication with relatives and friends, read news, search information, entertainment, purchase, and more. At the same time, developers, researchers, and students create new applications, services and solutions where the communication channel is through Internet. For example, social network sites, e-cloud and Internet of Things (IoT) represent the tendency in technology, which can be used in different scenarios. Other example of the relevance of Internet is the concept of Big Data where the behaviour of Internet users is collected and analysed to discover information. As a result, every day increases the number of Internet users around the world, making Internet part of our lives whether we agree or not.

In many of electronic services and web applications, Internet users need to fill out a web registration form. A web registration form is used for requesting personal

© Springer International Publishing AG 2017
J. Mejia et al. (eds.), *Trends and Applications in Software Engineering*, Advances in Intelligent Systems and Computing 537, DOI 10.1007/978-3-319-48523-2_23

information, such as address, age, name, sex, among others, to each Internet user [1]. The information is sent over the Internet from user's computer to the server. Then, the personal information is stored in a database and the system creates a new account. After that, each Internet user can access to the system using its account. This process was used since the beginning of Internet. According to Luke Wroblewski [2] a web registration form is:

"Registration forms are the gatekeepers to community membership", "Data input forms allow user to contribute or share information", and "Web forms are often the last and most important mile in a long journey."

This means that, Internet users share information with *n* companies when click submit button in order to create an account. The problem with the process is that Internet users disclose information under pressure [3]. This action has repercussion on privacy and security issues because personal information is used and collected by third parties [1].

On the other hand, young Internet users have integrated social network sites into their daily lives. According to [4] a social network site is:

"a web-based service that allow individual to 1) construct a public or semi-public profile within a bounded system, 2) articulate a list of other users with whom they share a connection, and 3) view and traverse their list of connections and those made by others within the system."

In this scenario, users of social network sites share their feelings, location, opinions, pictures, and special moments with others [4] and [5]. Moreover, the web service stores all the data in a database for certain time to offer services and promotions based on your preferences [6]. As a result, we can know many information about a person without meeting in the real world.

As a consequence of the use of web registration form and social network sites, organized crime and criminals can obtain several information about unprotected/vulnerable people. For example, when we click the submit button, without review the digital certificate or verify the legitimacy of the web site, our personal or financial information could be theft. This action is possible because many people do not know the relevance of security, in specific the digital certificate. Moreover, people share information through social network sites, forgetting their privacy.

Specifically, the contribution of this paper is double. First, we explain the security issues of web registration form, which mainly focuses on its vulnerabilities and its repercussion on the real life of users. Secondly, we collect data to allow an empirical investigation of the lack of users' privacy in social network sites. This paper exposes the drawbacks of web registration form and social network sites in order to find a new and more secure mechanisms to prevent or reduce attacks.

This paper is organized as follows. Section 2 presents a literature review of the users' perception about security and privacy in: Internet, e-commerce, virtual world, and social network sites. In Section 3, the security issues with the use of web registration form are described. Section 4 presents the security and privacy issues found in social network sites. Conclusions are given in Section 5.

2 Related Works

At the beginning, people found more disadvantages on Internet based on the online privacy and security risks [7]. This perception was founded on the cyber-attacks to business, damage caused by virus, and the absence of law [7]. Since that, some researchers have studied the Internet users' perception of privacy and security in three scenarios – electronic commerce, virtual worlds and social network sites [7], [8], [9] founding the same results.

The first examination of Internet user's perception of privacy and security was conducted by Kermek and Bubas [7]. They investigated the effect of privacy and security by comparing Internet and mass media among people. Their results found that Internet was viewed more unsecure in comparison with television and press. Two years later, Liebermann and Stashevsky [9] investigated the people's perception of risks as barriers to adopt e-commerce. They found that internet credit card stealing and supplying personal information were perceived as risk by people. In this way, Belanger, Hiller and Smith [10] explained how the privacy and security are two key components by Internet users when they want to purchase something in Internet. Moreover, Wang *et al.* [11] explained the relation among privacy and security as a positive effect in users of electronic bank. Later, Scott [12] studied the perception of e-business risk demonstrating that privacy and security still being perceived as risk. More recently, Mekovec and Hutinski [13] carried out an investigation about the Internet users' perception of privacy and security during their online activities. They concluded that trust in electronic commerce is based on privacy and security.

With the emergence of virtual worlds, new paradigms appeared from the field of psychology, anthropology and information technology. In this new scenario, where Internet users have the option to create a second life [14], privacy and security are perceived as risk [15]. Moreover, users identified the following risks: identity theft, frauds and virtual crime [1].

In this way, social network sites have opened a new door for researcher to study the relation among people [5] and its impact in different topic, such as: friendship, privacy, security, popularity. In terms of privacy, the studies of [15], [16] have evaluated the information disclosed in social network sites and security issues. In [16] the authors demonstrated that users share their personal data and they did not change the default privacy preferences. In order to identify the information disclosed by users in different social network sites, [1] carried out a study of demographic variables, usage context and usage patterns, finding that users disclosed their real name, picture of user, date of birth, and network of friends to unknown persons, while their phone number, physical address, instant messaging cont., and website are disclosed to friends. In [17] the authors found a correlation between users' perception of trust in the social network site and the information disclosed, getting similar results that [1]. Later, in [18] the authors described that users consider more personal sharing contact information than personal information. More recently, in [19] the authors found that behaviour must be protected as contact- and personal-information.

In this paper, we explain the security and privacy issues related with the use of web registration form and social network sites, exploring their impact on the real life of users. We evaluated the registration process of 37 web sites in terms of security and privacy.

3 Web Registration Form: Security and Privacy Issues

First, we describe the general process to create an account based on the use of web registration form. Fig. 1 shows the process carried out between Internet users and web servers to create an account. Then, we explain the method used to analyse the web registration form. Later, we explain the security and privacy issues found in this phase of our research.

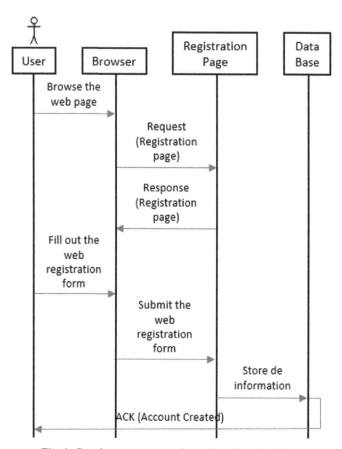

Fig. 1. Creating an account using web registration form.

The steps are as follows:

Step 1: Internet users want to create a new account. This action is based on users' interest.

Step 2: Internet users go to the web page. Users require a computer or mobile device and Internet connection to connect with the web server.

Step 3: Web browser displays the web registration form. The web server sends the registration form through a secure channel. The connection between the web server and users' computers should be secure through SSL (Secure Socket Layer) or TLS (Transport Layer Security) protocols.

Step 4: Internet users set up the web registration form. Users fill out the web registration form typing common information. The web server carries out the inline verification of the web registration form.

Step 5: Internet users submit the web registration form. Users send the web registration form through a secure channel, in most of the cases. The web server verifies the uniqueness of each ID or identity. If the verification fails, the user needs to rewrite a new one.

Step 6: Web server stores the information in a database. The web server stores the information of each user in a database.

Step 7: Web server creates a new account. The result of the process is the creation of a new account, which the web server and Internet user knows.

3.1 Method

This part of the research was conducted as follows: 1) we have evaluated the security of some web sites, considering its security and privacy during the registration phase and 2) we have created accounts in different web sites using fake information. Table 1 summarize the information obtained in this step.

The web sites evaluated include different classifications, giving us a general vision of security and privacy. All the web sites evaluated are in Mexico. We focused our attention in the use of digital certificate and security protection during the connection between the web server and web browser.

Then, we created some accounts in different web sites to know if we can get an account. Unfortunately, we have created a new account in web sites using fake information. In this study, fake information means that the information used to fill out the registration form is not genuine.

Table 1. Security and privacy evaluation of web sites.

Classifications	Digital certificate		Detection of fake information		Total web sites evaluated
	Yes	No	Yes	No	
Manufacturers	2			2	2
Public services	2	4	1	5	6
Real-estate	0	5		5	5
Service businesses	14	6		20	20
Transportation	2	2		4	4
	20	17	1	36	37

3.2 Results

As a result of our work, we found the following possible problems.

Problem 1: during the step 3, many web servers do not use digital certificate and security protocols to protect the users' information. This action represents a security risk because eavesdroppers can obtain such information.

Problem 2: during the step 4, the web server carries out the inline verification of the web registration form to ensure every single textbox is completed. However, web server does not verifies the legitimacy of the user's information.

Problem 3: each new account or new purchase online represents that Internet users share their personal and/or financial information with different web servers across the Internet. After that, users lose the control over such information because they do not manage the web server or database. Although, some countries have privacy law many others not.

Problem 4: in the case of social network sites, the managers of the system do not have any information about the criminal antecedents. As a result, criminals can get access to personal information of many users.

Problem 5: the personal and financial information stored in a database must be protected by the provider of the system. However, how we can be sure of that?

Problem 6: identity theft is a big problem for Internet users due to the following reasons: the difficulty to find a fake profile and the fake profile can affect the reputation of the victim.

Problem 7: when an Internet crime occurs, the police have problems to trace the real identity of the criminal from the virtual profile used to commit the attack.

Problem 8: in some cases, children or young people can get access to adult material because the system cannot verify the real age of the user.

As a final notes: 1) the web registration form remains the same when other applications have changed in the last years and 2) every day, many Internet users need to fill out a web registration form either to create a new account or to finalize a purchase online. This action represents, for Internet users, a tedious task, which can give as result that users decided to abandon the action.

4 Social Network Sites: Security and Privacy Issues

Social network sites are a global phenomenon which provides many advantages to users. However, users do not know or understand the security and privacy issues related with the use of social network sites, as we explain in this section. The problems were detected by means of the data collected in the survey and the experience obtained through an attack.

In previous works, the authors have explored the users' perception of security and privacy, and its consequences. However, their information were obtained from surveys and they analysed the data by means of statistical analysis. In our case, we wanted to know the vulnerabilities of the social network sites. For that reason, we carried out an attack creating a fake account.

4.1 Method

An online survey was devised to investigate the users' level of privacy in their account. The survey was published in our wall of Facebook waiting that our friends and friends of friends collaborated with the research. A total of 69 responses were received. In this part of the research, we wanted to know users' perception of privacy in their accounts, considering if users accept as friend unknown people and if users share pictures with relative and friends. In this way, we asked if their personal information is public.

The online survey consisted of 19 multiple-choice questions. The respondents shared based information related with their account of Facebook, including the amount of time with an account (1 year, 2-4 years, greater than 4 years), the knowledge of the data policy of Facebook (yes, no), and their age (younger or adult). Moreover, respondents specified if their account is public (yes, no, I don't know), if their share pictures with relatives and friends in their profile picture (only me, relatives and me, other), and if their have unknown people as friends (yes, no). Furthermore, they also indicated the name used (real name, nickname).

After that, we created an account using fake information and we created a social network with other students of the Universidad Autónoma de Ciudad Juárez. As a result, we were accepted as a member of a group and established a relationship with unknown students. By means of that, we had access to their accounts including pictures, friends, comments, and personal information.

4.2 Findings

The survey participants were $n = 69$, finding that 85.52% of the respondents used their real name and 81.16% of the respondents used their picture in Facebook profile. In this way, 75.36% of the participants share their hometown, work information, and hobbies. At this point, our results corroborate previous studies and characterize the Mexican user profile.

As a new result, we found that 23.19% of the respondents have as friend unknown people. Moreover, 88.41% of the respondents do not know if their account is public or not.

4.3 Attack

From our previous research, we decided to know the security mechanism used by the most popular social network site in Mexico, during the registration phase. In the first step, we created an account using fake information. As we expected, the account was created without problems (see Fig. 2-a. Then, we completed the profile – gender, relationship status, birthday, hometown, personal interests, education information, and work information – in order to give confidence to others (see Fig. 2-b).

In the second step, we sent many friend request to students of the Universidad Autónoma de Ciudad Juárez, receiving some confirmations. Then, we tried to be member of Facebook groups. Our fake account was accepted as member of some groups (see Fig. 2-c). At this point, users cannot know if the personal information of each user is true or false.

(a)   (b)   (c)

Fig. 2. Creating a fake account.

4.4 Results

As a result of our work, we found the following possible problems.

Problem 1: social network sites do not have security mechanisms to verify the personal information submitted by users. As a result, organized crime, criminals, and malicious users can create fake accounts.

Problem 2: users do not have a good education in terms of privacy. Users share many information in their wall – feeling, comments, gossip, pictures, and more – without worrying about who or whom can see it.

Problem 3: users do not respect the privacy of friends when they publish a picture. This action affect directly the privacy of each person who appears in the picture.

Problem 4: the police cannot trace easy and fast the personal information of the attacker when a security issue appears.

Problem 5: many accounts can be created using personal information of different person. Identity theft is a big security problem in social network sites.

5 Conclusions and Future Work

In this paper, we have evaluated the security and privacy implications on the use of web registration form and social network sites, corroborating previous results. Moreover, we have demonstrated that web registration form is the weakest link of security chain because the information is not verified by any mechanism. The web registration form is used in the registration process by all web services, which it means that, organized crime, criminal and malicious users can obtain personal and financial information of victims. Furthermore, attackers can create accounts using fake or theft information. As a result, identity theft and spoofing are a big problem for Internet users and police.

In this way, when Internet users sent their personal or financial information to web servers, they lose the control of such information. For example, if the database administrator want to copy their information to other storage device, the Internet users do not have knowledge about that.

Nowadays, we have designed a new survey based on previous works and we going to apply it using our personal account in social network sites and using the communication channels of the Universidad de la Salle Bajío, from September 1st to October 28th. We want to increase the number of participants and we want to have data from different type of Internet users.

Acknowledgments. This research was partially supported by La Universidad de la Salle Bajío. We thank the anonymous reviewers for their careful reading of our manuscript and their many insightful comments and suggestions, which significantly contributed to improve the quality of this paper.

References

1. Schrammel, J. Köffel, C., Tscheligi, M.: How Much do You Tell? Information Disclosure Behaviour in Different Types of Online Communities. In: Proceedings of the fourth international conference on Communities and technologies, pp. 275-284, ACM, New York (2009).
2. Wroblewski, L.: Web Form Design: Filling in the Blanks.Rosenfeld Media, (2008).
3. Joinson, A.N., Reips, U.-d., Buchanan, T., Paine Schofield, C.B. Human-Computer Interaction. 25, 1-24 (2010).
4. Boyd, D.M., Ellison, N.B.: Social Network Sites: Definition, History, and Scholarship. Journal of Computer-Mediated Communication. 13(1), 210-230 (2007).
5. Ellison, N.B., Steinfield, C., Lampe, C.: The Benefits of Facebook "Friends:" Social Capital and College Students' Use of Online Social Network Sites. Journal of Computer-Mediated Communication. 12(4), 1143-1168 (2007).
6. Facebook, https://www.facebook.com/policy.php
7. Kermek, D., Bubas, G.: Being connected to the Internet: Issues of privacy and security. Journal of Information and Organizational Sciences. 24(1), 55-67 (2000).
8. Stockton, R., Cunningham, S.: An Investigation into User Perceptions of Privacy and Trust and their Real-World Practices. In: Proceedings of the Tenth International Network Conference, pp. 139-149, Pymouth University, United Kingdom (2014).
9. Liebermann, Y., Stashevsky, S.: Perceived risks as barriers to Internet and e-commerce usage. Qualitative Market Research: An International Journal, 5(4), 291-300 (2002).
10. Belanger, F., Miller. J.S., Smith, W.J.: Trustworthiness in electronic commerce: the role of privacy, security and site attributes. The Journal of Strategic Information Systems, 11(3-4), 245-270 (2002).
11. Wang, Y.S., Wang, Y.M., Lin, H.H., Tang, T.I.: Determinants of user acceptance of Internet banking: an empirical study. International Journal of Service Industry Management, 14(5), 501-519 (2003).
12. Scott, J.E.: Measuring dimensions of perceived e-business risks. Information Systems and e-Business Management, 2(1), 31-55 (2004).
13. Mekovec, R, Hutinski, Z.: The role of perceived privacy and perceived security in online market. In: Proceedings of the 35th International Convention of Information

Communication Technology, Electronics and Microelectronics, pp. 1549-1554, IEEE, Opatija (2012).
14. Mennecke, B.E., McNeill, D., Ganis, M., Roche, E.M., Bray, D., Konsynski, B., Townsend, A.M., Lester, J.: Second Life and Other Virtual Worlds: A Roadmap for Research. Communications of the Association for Information Systems, 22, 371-388 (2008).
15. Leitch, S., Warren, M.: Security Issues Challenging Facebook. In: Proceedings of the 7th Australian Information Security Management Conference, pp. 137-142, (2009).
16. Gross, R., Acquisti, A.: Information Revelation and Privacy in Online Social Networks. In: Proceedings of the Workshop on Privacy in the Electronic Society, pp. 71-80, ACM, New York (2005.)
17. Dwyer, C., Hiltz, S.R., Passerini, K.: Trust and Privacy Concern with Social Networking Sites: A Comparison of Facebook and MySpace. In: Proceedings of 13th Americas Conferences on Information Systems, (2007).
18. Sahinaslan, E., Kantürk, A.: Information security awareness and awareness creation methods in organizations. In: Proceedings of the 11th Acaemic Information Conference, pp. 597-602, Sanliurfa (2009).
19. Ögütcü, G., Testik, Ö.M., Chouseinoglou, O.: Analysis of personal information security behavior and awareness. Computers & Security, 56, 83-93 (2015).

Determinant Factors in Post-Implementation Phase of ERP Systems

Jakelina Camizan Lozano, Sussy BayonaOré

Unidad de Posgrado de la Facultad de Ingeniería de Sistemas e Informática, Universidad Nacional Mayor de San Marcos (UNMSM). Av. Germán Amézaga s/n, Lima, Perú
jaquelina.camizan@unmsm.edu.pe, sbayonao@hotmail.com

Abstract. Implementing an ERP system, usually leads to a long-lasting and expensive project with several particularities that mostly have as consequences fails which leads to the failure of that implementation. That explains the importance of evaluating the success of an ERP project in the post-implementation phase and thus know how the ERP project has achieved their initial defined objectives. In this article is being presented a systematic revision of the researches made to the post-implementation phase of an ERP project with the purpose to investigate and identify the factors that have contributed to the success of those systems. A total of 73 articles of a 10-year period (2004-2015) were selected. The factors that helps to the early mentioned are a good project management, having well defined objectives from the beginning and the proper training of work teams.

Keywords: ERP Systems, Post-implementation factors, ERP benefits

1 Introduction

Nowadays, there is need to have a global vision of the organization and its behaviour in order to take strategic decisions [1]. The most important information system projects are the projects for the implementation called ERP systems (Enterprise Resource Planning) because of the extend time of implementation, a sizeable work team, their expensive cost and also the redesign of their processes, and once concluded presents a high chance of failure as consequence of the lack of confidence in the system from users' part.

According to studies made about the post-implementation phase of an ERP system [2], this process is successful only when the system is aligned to the business processes of the company and the strategy followed [3]. If there is not a correct alignment among those, the enterprise may suffer dangerous consequences. For that, the post-implementation phase of the ERP systems has become a critical topic of investigation, fact that is reflected by the increasing number of investigations that studies it and the increment in the generated business volume [4]. The cost to update an ERP involves between 25% and 33% of the total cost of the system [5] and the ERP projects have a high rate of failure. Likewise, there are various factors which contribute to the success of the ERP system projects, which needs to be identified. In this article a revision of the main studies about the post-implementation phase of the

© Springer International Publishing AG 2017   257
J. Mejia et al. (eds.), *Trends and Applications in Software Engineering*, Advances in Intelligent Systems and Computing 537, DOI 10.1007/978-3-319-48523-2_24

ERP system is presented in order to help professional researchers about the factors that has to be taken into account in the post implementation phase of an ERP system as well as the way to achieve the objectives defined at the beginning of the project. The results shown in this article provides a guide about the importance to have a successful Post-Implementation Phase.

Then this article depicts it in the next sections. Section 2 is an introduction to the ERP systems concepts, in which it is explained the differences between the implementation and post-implementation phases. In Section 3 is presented the method used to conduct the systematic revision based on Kitchenham's theory. Section 4 shows the results of the systematic review. Section 5 presents the discussion of the results. And, finally, Section 6 shows the conclusions.

2 Background

2.1. Enterprise Resource Planning Systems (ERP)

An ERP is an information system that integrates business processes with the purpose to create value and reduce costs while making available the right information to the right people at the right time, and that way help them to take the right decisions in order to handle the resources of an organization in a productive and proactive manner [6]. ERP systems are characterized to be integrated because they control all the processes in the organization related among each other, modular and adaptable.

2.2.   Difference Between Pre-implementation and Post-implementation

The life cycle of an ERP system can be understood in two big phases or "waves", differentiated by a central moment between them. In the first wave include the components of the ERP that the enterprise requires are selected, their adaption to the specific needs and after that implemented, ending with go live. The second wave is the post implementation phase, where it has to focus on supporting and enabling the organization to maximize the value and the return of investments made in the ERP system [7].

2.3. Critical Success Factors to the Post-Implementation Phase

We can name two categories of the factors to determine the post-implementation phase: the effort for continuous improvement which is associated to the improvement and integration of the system's processes, and the organizational support which is related to technical capabilities and soft skills of the work team and the ERP systems users [8].

3. Research Method

According to Kitchenham and Charters, a systematic revision seeks to identify, to evaluate and support all the relevant studies now available for a specific research question, a thematic area, a phenomenon of interest. Researches are executed properly and the data is retrieved. In the stage of extraction, the data is validated and is classified based on their "quality" so finally, results and conclusion are obtained [9]. The definition of a protocol is important and required to reduce the possibility of search bias due to the fact that the protocol specifies the methods used to guide the systematic revision. A summary of the protocol used in this revision is been given in the following sections.

3.1. Research Questions

The current systematic revision intends to answer to the following research questions (RQ):

- *RQ1: Which are the models or the frameworks that have been defined for the success of the post-implementation phase of ERP systems?*
- *RQ2: Which are the important factors in order to achieve the success in the post-implementation phase of ERP systems?*
- *RQ3: Which are the critical factors in the implementation phase of an ERP system that enables positive results in the post-implementation phase?*

3.2. Data Sources

The research strategy to find out the articles, has included automatic search into electronic data bases, and that way ensure that the main number of articles have been visited, even if this have caused redundancy in the results. The automatic search includes the most relevant indexes of the scientific mechanism studies, and a search of PDF files on the web pages of different academic sites like Staples and Niazi suggests [10]. The following electronic database were consulted: ACM Digital Library, Research Gate, AIS Electronic Library, Dialnet; ScienceDirect; ERP – PDF Documents (Google Scholar).

3.3. Search Terms

The search made in the electronic data bases derived from the key words contained in the research question. The strategy used to build up the search string was: 1) derive keywords from the research words, 2) Identify synonym or related ones to the keywords; 3) Grouping synonyms and related words with "O" identifier; 4) Group of each set of terms with "Y"; 5) Set the words in English as well in Spanish. In Table 1 the search strings used are presented.

Table 1. Search strings used

Search string	Conjunction
"ERP" or "Enterprise Resource Planning" or "ERP System" or "sistemas ERP"	AND
"postimplementation" or "post-implementación" or "models" or "modelos" or "methodological proposal"	AND
"successful" or "factors success" or "factors" or "critical factors of sucess".	AND

3.4. Criteria to Select an Article

The articles of this revision were selected based on the inclusion and exclusion criteria. In the Table 2 criteria of Inclusion and Exclusion are presented.

Table 2. Inclusion and Exclusion Criteria

Inclusion Criteria	Exclusion Criteria
Academic articles.	Articles that do not derives from an academically
Articles that defines the models used in	search engine.
the post-implementation phase of an	Articles not focused on ERP systems.
ERP.	Articles that are focused on methodologies for
Articles where there are defined the	implementation of ERP systems.
critical factors of success in the pre and	Articles based on experts' reviews without defining a
post implementation phase of an ERP.	specific experience on topic.
Written articles in English and Spanish.	Articles that focuses on the selection process of ERP
Articles published since 2004.	system.

3.5. The Procedure to Select an Article

To select an article, it relied on 4 phases in order to obtain a set of primary articles. Fig. 1shows the procedure divided on phases.

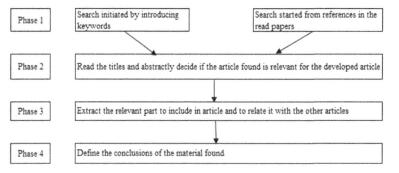

Fig. 1. Stages to select an article

To evaluate the quality of this revision has been used 11 criteria established by Dyba and Dingsoyr [11]. Those criteria are the following:
1) Is the paper based on an investigation (or simply is a "lesson learned" a report based on expert opinion)?, 2) There is a clear exposure of the research objectives?, 3) There is a proper depiction of the context in which the research was conducted?, 4) The design of the investigation is proper to face the objectives of the research?, 5) Was suitable the recruiting strategy to the aims of research?, 6) There was a control group with which they could compare how to structure?, 7) Was the data retrieved in a way that is related to the research topic?, 8) Data analysis were sufficiently rigorous?, 9) The relationship between the researcher and participants was considered to an appropriate level?, 10) There is a clear statement on the results?, 11) It is the value of the study for research or practice.

3.6 Data Synthesis

Results oriented for all articles found in order to: identify research criteria, the main topics that have been addressed and make a comparative analysis of the characteristics of the article, using demographic data, the timing, the type of methodology and defined topics, answers will be seeking to the research questions.

3.7 Development of the Revision

The revision begun with an automatic search and followed by a manual search, and that way there were identified the potential relevant items, and there were applied criteria of inclusion / exclusion. The search process, the search for final tests lasted from July 2014 to November 2015, and included the use of previously defined search strings.

3.7.1. Automatic Search

The first automatic search tests have begun in July 2014. In some engines, the search strings had to adapt, without losing its primary meaning and scope. After the search was completed, the results were recorded in a spreadsheet in order to facilitate the subsequent phase to identify potentially relevant articles. In November of 2015, it has been done the final search in the engines in order to include the articles published in 2015.

Table 3. Articles based on data sources

Data Source	No. of Articles	No. of Selected Articles
IEEE	12	3
Science Direct	102	21
Research Gate	88	11
Dialnet	4	1
AIS Electronic Library	25	6
Google Scholar	883	31
Total	**1114**	**73**

3.7.2. Potentially Relevant Articles:

The articles have been sorted by title in order to avoid redundancy. The articles for which the title, author (s), year and abstract were identical, were consider redundant. After the exclusion of the redundant elements, 73 results remained. The titles and summaries of the articles, resulted from the automatic search, have been read in order to identify the potentially relevant articles. Some articles were from medical or psychological or chemistry area, due to the fact that those specialties also uses the acronym "ERP" (Event-Related Potential). 73 articles in total were reviewed through the following procedure: 1) Read the summary, keywords, references and more information in the site that the search engine refers to, 2) There were verified if the cited references by the included articles, may be relevant to our investigation and purpose, and 3) There were checked if the same authors have published another potentially relevant articles.

In general, were excluded the articles that compare the results obtained by adopting an ERP and the relevant points to consider in another phase that is not post-implementation. This because these issues are outside of this research, which investigates the main factors for success of ERP systems in the post-implementation phase. Eight relevant articles were found by using this procedure of automatic search with keywords. This reinforces the importance of the technique to review the references in the previously identify documents with the purpose to find another relevant article.

4. Results

4.1 Trend Studies

The Figure 2 shows the quantity of articles by period in which it can be highlighted the increase in the number of articles that are published about the post-implementation phase in ERP systems since 2004 until 2015.

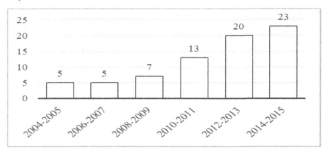

Fig. 2. Articles by period

Most articles have as main topics the following subjects:

• ERP: how it is managed during the last stage of implementation or during its production output.
• Post Implementation: it was considered all the aspects that impact on the post-implementation phase of a project of an ERP system.
• Critical factors for success in the pre-implementation phase: it was considered the articles that describe the important aspects for successful production output and in some cases serves as input in the post-implementation phase.
• Critical factors for success in the post-implementation phase: it was considered the articles which relate to the stable maintenance of ERP systems as well the models that were investigated related to the critical factors of success.

4.2 Study Types

From 73 identified articles, 42 articles are considered Empirical (58%), 15 articles are Theoretical (21%) and 16 articles are experience reports (22%). Those studies were developed in different environments; and there are projects that were developed in large, medium and small companies to ones that were derived from academic experiences. From the results presented above we can conclude that the empirical studies are the most important due to the fact their results are applied in different theoretical case studies.

5. Discussion

RQ1: Which are the models or the frameworks that have been defined for the success of the post-implementation phase of ERP systems?

Many of the information system models for the post-implementation phase were based on the Delone and McLean from 1992 where they considered the following aspects as part of the model: System and Information Quality, Use of the system, User satisfaction, Individual and Organizational Impact. This model has updates made by different authors since 1997 until 2007 where the last updates was in 2014. Ahad Zare Ravasan combined the provider and quality consultant criteria by calling them Inter-Organizational Impact and considers again the User because it is a vital point for the ERP to gain value and to be functional, besides the fact that is considered as a measure to know the success of an ERP system in the post-implementation phase. In 2015, Ahad Zare incorporates the workgroup impact and user satisfaction is replaced with the workgroup impact. Since 2013, from other criteria emerged types of models and frameworks on different aspects like there are project management, process reengineering, success management, organizational and individual level, stakeholders, conflicts and financial rates. Other types of frameworks are found where it is studied the relationship between the usability of the ERP systems and the benefits they generate through the knowledge integration mechanism, measure the performance of an ERP system based on the usability and the defined requirements. There is also models [12] related to measuring the failure of an ERP implementation based on Technological, Organizational and Environmental Theory (TOE) where it is considered the organizational, operational-technique aspects, the high management and the user inefficiency.

RQ2: Which are the important factors in order to achieve the success in the post-implementation phase of ERP systems?

Among the found factors in the revised articles (see Table 3) there are mainly considered the following: 1) Continuous improvement efforts in various processes that are part of the ERP system, 2) Organizational support that involves top management support, the ability of the internal ERP team, ongoing user training, communication between different areas of the organization, 3) Identify the risks about structure changes, high turnover of the ERP maintenance team, ERP code quality that belong to the post-implementation phase, 4) There is a revised article that contains factors that leads to the user acceptance of the system under the Technology Acceptance Model, this study is based on how the influence of 13 external factors increases the degree of user acceptance for the ERP system, 5) Internal and external technical conflicts that come from an understanding on how the conflict occurs during the ERP implementation in the organization change process, 6) With a proper handling of the presented factors for better or reduction of errors in the post-implementation phase, operational and management benefits can be obtained.

Table 3. Factors to achieve the success in the post-implementation phase of ERP systems.

Factor	Articles	Percentage
Success Measurement	20	21.98%
Organizational Changes	17	18.68%
Risk	10	10.99%
Use and Value	8	8.79%
Learning and Knowledge	6	6.59%
C.F. Implementation enables positive results to C.F. Post Impl.	6	6.59%
Change Management	5	5.49%
Financial Performance	4	4.40%
Benefits Post-implementation	4	4.40%
Projects Management	4	4.40%
Technological, Organizational and Environmental Theory	3	3.30%
Factors in organizational level	2	2.20%
Technology Acceptance Model (TAM)	2	2.20%

C.F: Critical Factors

RQ3: Which are the critical factors in the implementation phase of an ERP system that enables positive results in the post-implementation phase?

Several factors determine the success of the implementation on an ERP system. A summary of the factors mentioned in the literature: 1. Well defined objectives project planning phase who are the start point to define what is to be accomplished. 2. Support from top management from the beginning the high management has to be involved and has to know how to commit the work team in achieve the ERP systems. 3. Change Management where define risks and changes that affects the project and to know how to solve them properly. 4. Good administration in project management based in a methodology and apply the processed accordingly to reality of the organization. 5. Team project optimization, set a multidisciplinary team involving key users of the processes. 6. Conflict management between individuals that are part of the work team, team leads have to follow and how to have a proper communication between them. 7. Proper infrastructure according with the type of the software bought and the size of the project. 8. User training with the objective to achieve a good user performance through using the system while it is in production. Also, the founded factors that contribute with the business process improvement which are organizational changes, IT Government and business processes generated by changes in post implementation.

6. Conclusions and Further Research

This article is being presented a systematic revision of the researches made to the post-implementation phase of an ERP project with the purpose to discover the factors that have contributed to the success of those systems. From 1114 identified articles were selected 73 articles. The results of the revision show that the number of articles has incremented from year 2004 to the year 2015. The number of included articles and the fact that the number of the articles published in the last year indicates that the debate about this phase of the ERP projects is current. Mostly the studies are empirical in 58%, considered the most important because apply the theory in studies cases. Critical success factors belonging to the implementation phase and influencing the benefits that can occur in the post-implementation phase of the ERP systems were identified. It is important to analyse the success at different levels in an organization from various perspectives in order to gain a fuller appreciation of the success of the ERP system. It is suggested that in the ERP projects is to be considered the possible complementarities in managing a set of critical factors of success rather than focusing on individuals. Therefore, more research studies in this area are strongly recommended. Those studies may reuse the founded factors in order to propose a model that will contribute to the achievement of the expected benefits in the ERP systems.

References

1. Angela Patricia, B.: Factores de éxito para la implementación de sistemas de Planificación de recursos empresariales ERP (2014).
2. Booz-Allen, Hamilton: Insights enterprise resource planning. Big money down the drain? (2000).
3. Botta-Genoulaz, V., Miller, P. A. y Grabot, B.: A survey on the recent research literature on ERP systems. Computer in Industry, Vol. 56, pp. 510-522 (2005).
4. Salmerón Silvera, J. L., López Vargas, C.: Modelo Bidimensional de Riesgos del Mantenimiento de sistemas integrados de gestión (ERP). Investigaciones Europeas de Dirección y Economía de la Empresa Vol. 16, N° 3, 2010, pp. 173-190 (2010).
5. Carlino, J., Nelson, S., & Smith, N.: AMR research predicts enterprise applications market will reach $78 billion by 2004. AMR Research (2000).
6. McGaughey, R., Gunasekaran, A: Enterprise rResource Planning (ERP): Past, Present and future. International Journal of Enterprise Information Systems, Volume 3, Issue 3 (2007).
7. Willis, T. H. and Willis-Brown, A. H: Extending the value of ERP. Industrial Management & Data Systems, Vol. 102 Iss: 1, pp. 35 – 38 (2002).
8. Lorenzana Huertas, J. F.: Propuesta Metodologica para la Gestion de Mejorase en los sistemas de informacion en la etapa de Post Implementacion de una herramienta ERP: Estudio de caso. Universidad Nacional de Colombia (2014).
9. Olmedilla Arregui, J.J..: Revisión Sistemática de Métricas de Diseño Orientado a Objetos. Universidad Politécnica de Madrid, Facultad.de Informática (2005).
10. Staples M., Niazi M.: Systematic review of organizational motivations for adopting CMM-based. SPI Inform. Softw. Technol. 50 (7–8), pp. 605–620 (2008).
11. T. Dybå, T. Dingsøyr, Empirical studies of agile software development: a systematic review, Inform. Softw. Technol. 50 (9–10) (2008) 833–859, http://dx.doi.org/10.1016/j.infsof.2008.01.006.
12. Soltan, E. K. H., Jusoh, A., Mardani, A., Bagheri, M. M.: Successful Enterprise Resource Planning Post-Implementation: Contributions of Technological Factors. Journal of Soft Computing and Decision Support Systems, 2(4), 17-25 (2015).

Analysis, Specification and Design of an e-Commerce Platform That Supports Live Product Customization

João Barreira[1], José Martins [1, 2], Ramiro Gonçalves [1,2], Frederico Branco [1,2], Manuel Perez Cota[3]

[1] University of Trás-os-Montes e Alto Douro, Vila Real, Portugal
[2] INESC TEC and UTAD, Vila Real, Portugal
[3] University of Vigo, Vigo, Spain
JF_Barreira@hotmail.com, {ramiro, jmartins, fbranco}@utad.pt, mpcota@uvigo.es

Abstract. In recent years, the demand from online customers has become a major problem, the variety of choice and the need for them to feel special triggered the desire to customize the products they want to buy. This high customer demand for customized products has been one of the biggest obstacles that companies have encountered. To achieve an adequate response to customer demands, companies need to adopt tools that offer to customers exactly what they want, in other words, allow customers to customize the products they want to buy. This research aims to perceive if using CMS platforms and low-cost software might be a proper solution for enterprises who intend to not only sell their products online, but also want their customers to be able to perform live product customization during their purchase.

Keywords: e-commerce; e-commerce platforms; product customization; CMS.

1. Introduction

In a time where firms tend to transform themselves in order to become the most innovative and distinctive, ICT (Information and Communication Technologies) have the ability to help firms expanding to new markets and trigger their customer's loyalty [1].
The network framework behind the Internet and all associated ICTs has led so significant changes, with consequent improvements, to existing electronic commerce initiatives, thus the existing interest demonstrated by organizations on developing better and more customer oriented e-commerce initiatives [2, 3].
The numbers behind e-commerce adoption are very significant [4] and the ability to access e-commerce platforms from multiple devices has helped trigger the users willingness towards these technologies [5, 6]. As argued by Fogliatto, et al. [7], as the number of adoption users increases, so does the need for more adjustable e-commerce platforms that allow users not only to purchase the wanted products but also to personalize them in order to adjust to their specific will or needs.
The development of e-commerce platforms to the specific needs or an organization can be, as presented by Kohan [8], a very expensive activity. This represents and even significant impact to the organization when one of the associated

© Springer International Publishing AG 2017
J. Mejia et al. (eds.), *Trends and Applications in Software Engineering*, Advances in Intelligent Systems and Computing 537, DOI 10.1007/978-3-319-48523-2_25

requirements is the development of real-time product customization in order for the platform users to be able to customize the products they want to purchase to their own taste or need. With this in mind, and assuming the adoption of content management systems (CMS) as the basis for the development of not only websites but also e-commerce and e-business platforms [9], we developed a research project directed at perceiving if (low-cost or free) CMS can be used to properly develop e-commerce platforms that allow live product customization.

2. Research objectives and methodology

As it was previously mentioned, customers high demand for customized products when using e-commerce platforms has been in the agenda of all major market players for the past couple of years [10, 11]. From the perspective of the research team behind this project, the solution that could help enterprises, particularly small and medium sized ones, is the use of CMS technologies, thus developing low cost and highly customizable e-commerce platforms in a short period of time.

Drawing on the above, the two main objectives behind this research project were the use of CMS technologies to develop an e-commerce platform prototype that allowed its users to easily customize the available products and, at the same time, achieved a set of guidelines on how enterprises could apply this approach.

The performed research was performed in several stages: 1) conception – identification of the research problem and inherent objectives; 2) theoretical approach – analysis of existing scientific and technical literature that focused on the same issues and that allowed to achieve a broader understanding of the state-of-the-art; 3) technology analysis – identification and analysis of existing CMS technologies that could be used to develop an e-commerce platform that allowed its customers to customized the available products; 4) prototype – specification and development of an e-commerce platform prototype that allows users to customize products; and 5) guideline – development of a set of guidelines that organizations can use in order to create their own customizable e-commerce platforms.

3. Conceptual Framework

In the Information and Knowledge society where we all live, the use of Internet related tools and technologies has become a daily routine, hence the immense number of users connected online that, during their daily life, search and buy products online [1, 12, 13]. This assumption has led enterprises (big and small) to invest a significant amount of effort and capital towards the development of e-commerce platforms that not only attract new customers but that also trigger existing customer's loyalty [14-16].

Implementing an online e-business initiative is not an easy task for enterprises. A focused and thorough analysis to the enterprise structure, to its business needs, to its market and to its mid to long-term goals is required. If, by performing the referred analysis, one of the achieved conclusions is that e-commerce is critical to the organization, than it must be considered with open mind and all necessary caution and planning [17].

According to Hartono, et al. [18] and Zhang, et al. [19] an e-commerce platform can be considered an information system abstraction that acts as a support for both organizations and end-users to perform transactions online. Conceptually, these platforms are directed at allowing users to trade products or services with each other without having the need to know each other.

As argued by Chaffey [20], in order to reach success, an e-commerce platform must comply with a set of primary requirements: a) allow for catalogue management; b) allow for payment methods management; c) allow for orders management; d) implement an optimized search engine; e) allow to create business analysis in order to keep track of all transactions that are being made; and f) allow for customers to customize products in order for it to be almost an individual statement instead of a mass product. Thus, and acknowledging the existing difficulties in developing reliable and adequate e-commerce platforms, one can assume that choosing what technology to use for support the firm's e-commerce initiative is an extremely critical task.

The choice of the technology that will support the e-commerce initiative must have in consideration that despite the chosen technology, it will be used to support a long-term initiative and that it should be scalable enough to allow for the e-commerce platform to grow [21]. With the current frantic growth of Internet use, existing digital content has also grown and, alongside this, enterprises started to need technological tools that allows them to properly manage all of the content they've created and to create new dynamic and adaptable content. One of the technological solutions firms have adopted hoping to solve this issue were web based Content Management Systems (CMS). This systems allow to easily collect, manage and publish Web content that can be accessed by all types of devices [21].

4. CMS e-Commerce Platform with Live Product Customization

As it was mentioned early in the manuscript, in order to build a successful e-commerce platform, one must take accurate and well supported decisions towards what software to use as the basis for the development activities. This is even truer when analysing this issue from the CMS modular perspective. The CMS modules to implement and how to perform the needed configurations are decisions that the team behind the e-commerce initiative must be prepared to take.

According to several authors [17, 22-24], in order for a CMS-based e-commerce platform to work properly and achieve the necessary level of success, it must implement the set of modules necessary for it to be aligned with both the enterprise business and the market needs and desires. Examples of these modules are the ones responsible for connecting the e-commerce platform with social networks sites (aiming on reaching new customers and informally collect important feedback on both the brand and the products/services), the modules responsible for the Multilanguage feature (allowing for users from several countries to use the platform and purchase whatever products or services are in sail, hence increasing the overall profit), and the modules that allow for a proper presentation of all product details and characteristics (given the user cannot hold the product, than all relevant information must be presented to the user in order for him to perform an informed purchase).

As a complement to the above, Carmona, et al. [25] and Deng and Poole [26] also indicate that e-commerce platform must be designed in order to attract customers and trigger their desire to search for products and to purchase. Hence, well written, simple and straightforward texts will help users to trust the platform, as simple and intuitive navigation patterns will help users to navigate through the platform without any type of friction.

For the focus of our research we have chosen Prestashop [27] to serve as the CMS that would support the development of the e-commerce platform prototype. This software is free to use, its code is open-source and there are over 300 already implemented features that are available directly from the installation. Prestashop has a set of functional areas that can be configured in order to improve the e-commerce platform interface, navigation, product catalogue management, orders management and client's management.

4.1.   Live Product Customization

As ICT usage grows and users become increasingly demanding on the level of customization associated with each product or service they purchase online, e-commerce platforms requirements also have to adjust. One solution to this issue is mass customization, however in order for a Prestashop-based e-commerce platform to be able to allow for its customers to customize the available products it must be integrated with a product customization module.

One of the most used Prestashop modules for product customization is "Product Customization 5.0" whose technical features and compatibility safeguards allow for a proper and without issues installation and configuration. This module allows for the definition of a set of product features and characteristics that give the user the capacity to chose those features and customize the product at his will.

Figure 1 – Example of a Prestashop product detail screen.

In Figure 1 one can visualize an example of a product detail screen where users can, not only chose his preference on the predefined set of options associated with the product, but also choose to further customize the product by clicking the "customize" option. By choosing to perform this advance customization, the user is directed to a new, and more complex, screen (Figure 2), where he can visualize the product and perform real-time customizations, such as adding images or texts to the product. The example shown in Figure 2 clearly shows all the customization options available to the platform users.

Figure 2 – Example of Prestashop product customization screen after installing "Product Customization 5.0" module.

As one can see on Figure 3, if the user chooses the option to add an image to the product he is customizing, then this new screen is shown and the platform gives the option to upload images and insert them over the product base image, hence allowing for a live customization action.

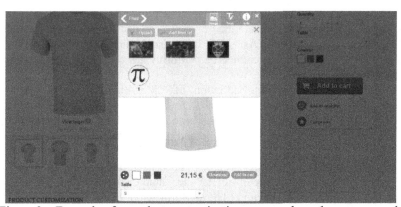

Figure 3 – Example of a product customization screen where the user can upload images and perform a live customization.

The product customization module offers another set of very important features for the accurate and straightforward execution of the live customization tasks. For instance, it allows adding multiple product images in order to increase the product customization range (Ex.: A t-shirt might be seen in four different positions – front, back, left and right – and users might want to customize each side).

As one can simple acknowledge, not all products are customizable, but despite that, the platform administrator has the option to activate customization features for all products.

4.2.   Platform Testing

The testing phase is one of the most important ones on the software development cycle [28, 29]. In what refers to Web applications and platforms testing, the main goal is not only to test the existence of errors or functional irregularities but also to test for the possible interaction and quality issues that might appear after users start to use the platforms. Therefore, one must have a considerable knowledge on the application architecture and all of its features and characteristics in order to accurately validate its quality. In order to properly validate a Web application there are several tests that one should perform: functional tests, data input tests, database test, user interface test, usability and accessibility tests, security tests and performance and scalability tests.

The perform of test is critical, given that it will help diminish the number of application and functional errors at the same pace that it triggers and enhances the software quality and performance. In order to properly test a software several profiles of users should be used, in order to not only access the more technical features but to also unveil possible issues that were never taken in consideration during the requirements specification and development stages.

5. Conclusions

For some years now e-commerce platforms users demand for highly personalized products has been increasing significantly. This issue tends to negatively impact enterprises who chose not to update their e-commerce platforms towards meeting the abovementioned requirement. As state by existing literature, this inability or unwillingness to upgrade the e-commerce platforms has causes such as the inherent cost and effort.

Considering the extrapolating adoption of CMS platforms to develop Web solutions (institutional websites, e-commerce and e-business platforms, etc.) a solution proposal for the referred issue has been suggested: an e-commerce platform built with a CMS that allows for a simple and quick customization of the available products.

The implemented prototype has been developed using Prestashop, an open-source e-commerce platform that uses mysql as the database engine and that has very reduced execution costs associated. The performed out-of-the-box installation and configuration of Prestashop, alongside with an installation of a product customization module (with a very reduced cost), has allowed to acknowledge that using this type of approach does not require a significant amount of technical knowledge and might be a fair an adequate solution for the majority of SMEs (Small and Medium Enterprises) that want to take the next step and start selling their products online.

References

[1]   R. Goncalves, S. Gomes, J. Martins, and C. Marques, "Electronic commerce as a competitive advantage: The SMEs from Trás-os-Montes e Alto Douro," in *2014 9th Iberian Conference on Information Systems and Technologies (CISTI)*, ed, 2014.

[2]   D. Campbell, J. Wells, and J. Valacich, "Breaking the ice in B2C relationships: Understanding pre-adoption e-commerce attraction," *Information Systems Research*, vol. 24, pp. 219-238, 2013.

[3]   I. Sila, "The state of empirical research on the adoption and diffusion of business-to-business e-commerce," *International Journal of Electronic Business*, vol. 12, pp. 258-301, 2015.

[4]   Ecommerce-Europe. (2015, 2016-03-05). *Global e-commerce turnover grew by 24.0% to reach $ 1,943bn in 2014*. Available: http://www.ecommerce-europe.eu/news/2015/global-e-commerce-turnover-grew-by-24.0-to-reach-1943bn-in-2014

[5]   H. Awa, O. Ojiabo, and B. Emecheta, "Integrating TAM, TPB and TOE frameworks and expanding their characteristic constructs for e-commerce adoption by SMEs," *Journal of Science & Technology Policy Management*, vol. 6, pp. 76-94, 2015.

[6]   L. Einav, J. Levin, I. Popov, and N. Sundaresan, "Growth, adoption, and use of mobile E-commerce," *The American economic review*, vol. 104, pp. 489-494, 2014.

[7]   F. Fogliatto, G. da Silveira, and D. Borenstein, "The mass customization decade: An updated review of the literature," *International Journal of Production Economics*, vol. 138, pp. 14-25, 2012.

[8]   B. Kohan. (2015, 2016-03-07). *Web Development Cost / Rate Comparison - Different Types of Custom Web Application Companies*. Available: http://www.comentum.com/web-development-cost-rate-comparison.html

[9]   A. Mirdha, A. Jain, and K. Shah, "Comparative analysis of open source content management systems," in *Computational Intelligence and Computing Research (ICCIC), 2014 IEEE International Conference on*, 2014, pp. 1-4.

[10]   E. Spaulding and C. Perry, "Making it personal: Rules for success in product customization," *Bain & Company Publication*, 2013.

[11]   F. Piller, T. Harzer, C. Ihl, and F. Salvador, "Strategic Capabilities of Mass Customization Based E-Commerce: Construct Development and Empirical Test," in *2014 47th Hawaii International Conference on System Sciences*, 2014, pp. 3255-3264.

[12]   J. Martins, R. Gonçalves, T. Oliveira, M. Cota, and F. Branco, "Understanding the determinants of social network sites adoption at firm level: A mixed methodology approach," *Electronic Commerce Research and Applications*, vol. 18, pp. 10-26, 2016.

[13]   J. Martins, R. Gonçalves, J. Pereira, and M. Cota, "Iberia 2.0: A way to leverage Web 2.0 in organizations," in *Information Systems and Technologies (CISTI), 2012 7th Iberian Conference on*, 2012, pp. 1-7.

[14]   C. Yoo, Y. Kim, and G. Sanders, "The impact of interactivity of electronic word of mouth systems and E-Quality on decision support in the context of the e-marketplace," *Information & Management*, vol. 52, pp. 496-505, 2015.

[15]   T. Oliveira and G. Dhillon, "From adoption to routinization of B2B e-Commerce: understanding patterns across Europe," *Journal of Global Information Management (JGIM)*, vol. 23, pp. 24-43, 2015.

[16]   R. Gonçalves, J. Martins, J. Pereira, M. Cota, and F. Branco, "Promoting e-Commerce Software Platforms Adoption as a Means to Overcome Domestic Crises: The Cases of Portugal and Spain Approached from a Focus-Group Perspective," in *Trends and Applications in Software Engineering: Proceedings of the 4th International Conference on Software Process Improvement CIMPS'2015*, J. Mejia, M. Munoz, Á. Rocha, and J. Calvo-Manzano, Eds., ed Cham: Springer International Publishing, 2016, pp. 259-269.

[17]   Z. Huang and M. Benyoucef, "From e-commerce to social commerce: A close look at design features," *Electronic Commerce Research and Applications*, vol. 12, pp. 246-259, 7// 2013.

[18]   E. Hartono, C. Holsapple, K. Kim, K. Na, and J. Simpson, "Measuring perceived security in B2C electronic commerce website usage: A respecification and validation," *Decision Support Systems*, vol. 62, pp. 11-21, 2014.

[19]   Y. Zhang, R. Mukherjee, and B. Soetarman, "Concept extraction and e-commerce applications," *Electronic Commerce Research and Applications*, vol. 12, pp. 289-296, 2013.

[20]   D. Chaffey, *E-business and E-commerce Management: Strategy, Implementation and Practice*: Pearson Education, 2007.

[21]   F. Trias, "Building CMS-based Web applications using a model-driven approach," in *2012 Sixth International Conference on Research Challenges in Information Science (RCIS)*, 2012, pp. 1-6.

[22]   A. Kankanhalli, O. Lee, and K. Lim, "Knowledge reuse through electronic repositories: A study in the context of customer service support," *Information & Management*, vol. 48, pp. 106-113, 2011.

[23]   A. Schlemmer and S. Padovani, "Estudo analítico dos estágios e informações que compõem a compra online," *InfoDesign: Revista Brasileira de Design da Informação*, vol. 10, 2013.

[24]   C. Kim, W. Tao, N. Shin, and K. Kim, "An empirical study of customers' perceptions of security and trust in e-payment systems," *Electronic Commerce Research and Applications*, vol. 9, pp. 84-95, 2010.

[25]   C. Carmona, S. Ramírez-Gallego, F. Torres, E. Bernal, M. del Jesús, and S. García, "Web usage mining to improve the design of an e-commerce website: OrOliveSur. com," *Expert Systems with Applications*, vol. 39, pp. 11243-11249, 2012.

[26]   L. Deng and M. Poole, "Aesthetic design of e-commerce web pages–Webpage Complexity, Order and preference," *Electronic Commerce Research and Applications*, vol. 11, pp. 420-440, 2012.

[27]   Prestashop, "Prestashop - Free e-commerce software," v1.6.1.6 ed, 2016.

[28]   P. Jorgensen, *Software testing: a craftsman's approach*: CRC press, 2016.

[29]   M. Mäntylä, B. Adams, F. Khomh, E. Engström, and K. Petersen, "On rapid releases and software testing: a case study and a semi-systematic literature review," *Empirical Software Engineering*, vol. 20, pp. 1384-1425, 2015.

Tourism Recommendation System based in User Functionality and Points-of-Interest Accessibility levels

Filipe Santos [1], Ana Almeida[1], Constantino Martins[1], Paulo Oliveira[2],
Ramiro Gonçalves[2]

[1] Institute of Engineering – Polytechnic of Porto, Computer Science Department,
Porto, Portugal
{jpe, amn, acm}@isep.ipp.pt
[2] University of Trás-os-Montes e Alto Douro, INESC TEC,
Vila Real, Portugal
{oliveira, ramiro}@utad.pt

Abstract. This paper describes a proposal to develop a Tourism Recommendation System based in Users and Points-of-Interest (POI) profiles The main focus of this work is to evaluate if gathered user's physical and psychological functionality levels will return more accurate recommendation results. This work also aims to contribute with a different way to classify (POI) considering their capacity measured in accessibility levels to receive tourists with certain levels of physical and psychological issues that will be described in this paper sections.

Keywords: tourism; recommendation; user profiles; points-of-interest, functionality, accessibility;

1 Introduction

User modeling implementation is traditionally performed recurring to two sets of techniques a: knowledge-based and behavioral [1]. Knowledge-based adaptation is normally the result of information gathered using forms, queries and other user studies, with the purpose to produce a set of heuristics. Behavioral adaptation is related with user monitoring during his daily tasks and activities. This work proposes a knowledge-based approach regarding the usage of information collected through forms. The main focus is to consider distinct user related information like relation between an user and a set of stereotypes defined by Tourism of Portugal (business, nature, sun and sea, etc.) and functionality levels in physical and psychological issues like locomotion, eyesight, acrophobia, agoraphobia and claustrophobia to create more accurate profiles. It is important to see user modeling as a process among others that compose an application. User modeling can be defined as a set of techniques that allow systems to retain users' information and that allow the use of this information in several ways with main purpose of improving and customizing user experience within that system [2]. It is normally a process that begins with a suitable representation of the user, or user model, which can be the addition of different components. Further, that information is used to infer and generate new knowledge that can either be added to the user model or used by the systems to adapt it, which allow system's efficiency

© Springer International Publishing AG 2017
J. Mejia et al. (eds.), *Trends and Applications in Software Engineering*, Advances in Intelligent Systems and Computing 537, DOI 10.1007/978-3-319-48523-2_26

enhancement, at least in the user perspective. This works aims to consider also a different approach to classify POI, where each one's accessibility levels related with the physical and psychological issues is evaluated.

A Recommender System can be defined like a collection of different techniques used by systems filter and organize its items in order to select either the best ones or the most suitable ones for presentation, according to the user [3]. Although the most common scenario is when the system has to choose the best items from a certain group which otherwise (without the filtering) would be randomly selected, there are other more important cases where certain items or types of items just can't be shown to the user at a given moment, for example, due to handicap issues. A complete recommender system should therefore be prepared to handle both types of situations. The mode of operation normally used by recommender systems is to use a knowledge base (the user model) as the basis for a series of calculations to infer which are going to be, amongst all the items available, the ones that will better please the user, according to a wide variety of theories or approaches. In this work is considered that the best way to please users is respecting their stereotype (nature, business, residential, etc.) relation accuracy and functionality levels (these levels are related with users abilities like locomotion, vision and phobias). The accessibility levels defined for each POI (it can be described as POI profile) is also a key difference in this work when compared with others.

Recommend something to someone carries an implicit responsibility to whom does that because it is important to assure accuracy and quality in the recommendation results. These systems are basically based in three types of paradigms (content, collaborative and knowledge-based) and all their possible combinations [4][5][6][7]. Content-based filtering tries to capture information from within the content of unstructured or unorganized item data elements, such as textual or descriptive attributes, generally including powerful text mining algorithms from the information retrieval area. Collaborative filtering (also called social-filtering) is one of the currently most used techniques and was greatly influenced by the Web 2.0 ("social web") phenomena. It relies on other user's information for recommending items to the current user [4]. Knowledge-based filtering is almost inevitable to use, because it means using any form of domain knowledge in a recommender system.

These two fields, user/POI modeling and recommender systems, are the core part of this work, in addition to a set of tools used to collect important data needed.

1.1 Motivation and Objectives

The main goal of this research is the development of a recommendation system that takes deeper user knowledge and is able to respect their unique preferences and functionality levels. These functionality levels are calculated regarding accessibility issues like locomotion, vision and phobias like acrophobia, agoraphobia and claustrophobia in addition to basic information like age, gender, nationality and other user preferences gathered by comparison with some standardized stereotypes.

This research assumes a vital importance in tourism recommendation systems and any other area where more accurate individual user knowledge is the key factor to obtain better results. The fulfillment of individual users needs in coexistence with the respect for their own physical and psychological limitations is probably one of most important objectives for this research work.

2 State of Art

2.1 User Profiling

In documents related with computer software history, it is generally explained that many software domains changed from a machine-perspective methodology, where the software itself performs the main role and the user is obligated to adapt itself, to a user-oriented development, where the software is created to fulfill user needs, objectives and desires. This was a similar movement to the occurred in industry strategies, that changed from a product-oriented approach to a customer-oriented paradigm, where all products are developed depending on market studies results. In computer systems, user needs performs a similar role [8] where users must have the feeling that the system works for them and not the contrary improving accuracy and efficiency of daily tasks. A costumer-oriented software will benefit from users' trust and confidence acquiring an important competitive advantage in an increased competition world. This is the part where user modeling can perform the difference. For instance in this work [9] related with education processes, authors model students as individuals, differ in their social, intellectual, physical, psychological, emotional and also ethnic characteristics in order to obtain better student's results.

Historically, first research's related with user modeling appears in literature in the 70's conducted by Allen, Choen, Perrault and Elaine Rich [1]. In fact, during this literature review, became clear that Rich [10] and more recently Kobsa [2] are two of the most important references related to this subject. In the last decades, several different systems were developed to store different kinds of user information. Some of those applications were analyzed and reviewed in works done by Morik, Kobsa, Wahlster and McTear in 2001. In those first systems, user modeling was an application part, which caused difficulties to separate users' profile related processes from the other application components. This was a normal problem in software design previous to of software encapsulation and modularization techniques has became popular. Despite the technology evolution related with users' profiles modeling (it has become more complex and intelligent, by making use of newly technological evolutions), the basic concepts and ideas and problems that turned possible the appearance of this research area are almost the same: the identification of user needs, desires, personalities and, most importantly, objectives.

In recent years different user modeling techniques and methodologies were used to represent knowledge, some of them are data representation oriented and others data inference oriented. The UM techniques (linear models, decision trees, neural networks, text mining, Bayesian networks and data mining) are all forms of predictive statistical models, since they are applied in areas with thousands or millions of items (from products, clients, actions, etc.) and can also beneficiate from recent machine learning evolution [11]. Finally, not all of them might actually be applied in some domains, due to their nature.

Linear models is one of most common techniques, and it can probably even be said that every system uses linear models, one way or another, although there are systems that entirely rely on linear models and explore all their possibilities. These models are easy to build and understand; they are efficient and assume probabilistic data as believable effects, which has been a successfully employed theory so far [10]. They generally use weighted sums or means of frequently accessed items to conclude user interests, in the case of the product applications described previously, and, therefore, infer the likelihood for new unknown items.

2.2 Recommendation Systems

In this section will be described some existing recommendation systems, according with their reference capabilities.

TIP [12] and Heracles [13] provides recommendation services through mobile devices for tourism. These services implement hybrid algorithms to calculate tourist preferences, using the defined tourist profile and location data (location-aware).

Proximo [14] is a location-aware mobile and recommendation system that fits the pure paradigm approach. It guides users through tours within buildings using Java and Bluetooth technologies. The mobile device also tracks the user location and builds a context, providing the system with important information. The user position is taken by "sniffing out" the fixed Bluetooth devices or low-cost beacons deployed in the area of use. Proximo pure collaborative recommendation system relies on its user's item ratings to provide recommendations.

In GeoNotes [15] system tries to blur the boundary between physical and digital space (ubiquitous computing and augmented reality). At the same time, it strives to socially enhance digital space (collaborative filtering, social navigation, etc.) by allowing users to participate in the creation of the information space. GeoNotes is a location-based information system that allows the user to access information in relation to the user's position in geographical space.

The tourism-oriented mobile GIS application MacauMap [16] designed for the city of Macau allows map navigation while displaying the user current location. It also provides information about the public bus network and bus guides for calculating optimal bus routes. It also provides sightseeing guides with information about museums, churches, temples, hotels, restaurants and other places of interest, along with their location on the map.

EtPlanner is a mobile planning assistant [17] that allows the creation of personalized tourism stays. Using a mobile device (e.g. a PDA or mobile phone) the costumer's stay is intelligently planned. This way the user can be assisted before, during and after his journey.

The personal mobile assistant mobiDENK [18] has been developed for a tour to the Herrenhausen Gardens in Hanover and includes POI on which historical information and images of the most significant features are presented on a PDA. It focuses on drawing the user's attention to historic sites and provides location-based multimedia information at different sightseeing spots while displaying the person's current location on a map.

Acesssights [19] is a subproject of mobiDENK, and is intended to provide tourist information to both normally sighted users and visually impaired people traveling in the Gardens. Normally sighted users will make use of both senses to obtain information and may simply follow a guide map, while blind people listen to information. The system uses loudness in order to point out the distance between the user's current location and point of interests, by simply making the voice signal get louder as the user comes closer to the point.

Tourist Guide [20] is a location based tourist guide application for the outdoor environment and it was developed for visitors to the Mawson Lakes campus (of the University of South Australia) and the North Terrace precinct in the Adelaide city center. The user interacts with the system using a PDA that displays his current position along with detailed information about specific nearby POI (a self guided tour of a specific area) like buildings, attractions and nearby utilities such as public telephones and toilets.

The application m-ToGuide [21] is targeted for the European tourism market and offers location-specific multimedia information about major monuments and POI. A portable, handheld terminal is used to exchange information between the m-ToGuide system and the tourist. All information and services delivered to the tourist will be relevant to his/her specific location (location-based) and tailored to that end-user's personal profile. The m-ToGuide experience can be personalized to give tourists direct access to the information and services they prefer.

The UK city of Lancaster has available the application The GUIDE [22] that provides city visitor's up-to-dated and context-aware hypermedia information while they explore the city.

Crumpet [23] provides new information delivery services for a far more heterogeneous tourist population. The services proposed by CRUMPET take advantage of integrating four key emerging technology domains and applying them to the tourism domain: location-aware services, personalized user interaction, seamlessly accessible multi-media mobile communication, and smart component-based middleware or "smartware" that uses Multi-Agent Technology. The system learns more about the user's preferences while he's traveling and interacting with the system itself.

Cyberguide system [24] was developed at the Georgia Institute of Technology (GIT), Atlanta, USA. It is based on the ubiquitous computing concept and focuses on mobile context-aware tour guide. The system was designed to assist a visitor in a tour to the GIT, and helps the user obtaining information about the demos in display. Knowledge of the user's current location, as well as a history of past locations are used to provide more of the kind of services that we come to expect from a real tour guide.

CATIS [25] is a context-aware tourist information system with a Web service-based architecture. The context elements considered to this project are location, time of day, speed, direction of travel and personal preferences. This system will provide the user with relevant information according to his location and the current time.

Deep Map application [26] realizes the vision of a future tourist guidance system that works as a mobile guide and as a web-based planning tool. It is a mobile system that aids tourists with navigating through the city of Heidelberg by generating personal guided tours. Such a tour shall consider personal interests and needs, social and cultural backgrounds (e.g. age, education and gender), type of transportation (e.g. car, foot, bike or wheelchair) and other circumstances from season, weather and traffic conditions, to time and financial resources.

In Tousplan project [27][28] the development of a Tours Planning Support System (TOURS PLAN) is proposed which intends to help tourists in finding a personalized tour plan allowing them to use their time efficiently and promote the culture and national tourism. Hence, this research focuses on tour planning support, aiming to at define and adapt a visit plan combining, in a tour, the most adequate tourism products, namely interesting places to visit, attractions, restaurants and accommodation, according to tourist's specific profile (which includes interests, personal values, wishes, constraints and disabilities) and available transportation modes between the selected products. Functioning schedules are considered as well as transportation schedules. This project tries to efficiently address the core of the tour planning process. Hence, it defines an optimization model that clearly represents the described tour-planning problem and designs a heuristic algorithm that effectively tackles that problem. The Traveling Salesman Problem (TSP) and some of its variations with additional constraints like time windows (TSPTW) or the Prize Collecting Travelling Salesman Problem (PCTSP) are used as basis for the development of algorithms that address tour-planning issues.

ITravel application [29] is based in ratings provided by other tourists with similar interests. The approach used employs mobile peer-to-peer communications for exchanging ratings via their mobile devices. Data exchange is based in wireless RF-communication technologies present in mobile devices that allow users to effectively share their ratings toward visited attractions.

3D-GIS Hybrid is a context-aware mobile recommender system whose goals are Ubiquity (users may use the system wherever they like using the mobile platform), Location-awareness (recommendations provided are adapted to the user's current location) and 3D-interface (it includes a 3D solution with innovative features as 3D geovisualization, location, etc.)[30].

POST-VIA 360 [31], is a platform devoted to support the whole life-cycle of tourism loyalty after the first visit that is designed to collect data from the initial visit by means of pervasive approaches. This data used to produce accurate after visit data and, once returned, is able to offer relevant recommendations based on positioning and bio-inspired recommender systems. The validation is based in a case study comparing recommendations from the application and a group of experts.

3   Proposed Approach

In this work a recommendation system based in users and POI profiles that will be linked with an algorithm to extract their accuracy relation is proposed (if this is high this POI should be recommended to this user). In Figure 1, a high level architecture that focused in the relation between user functionality and POI accessibility is shown.

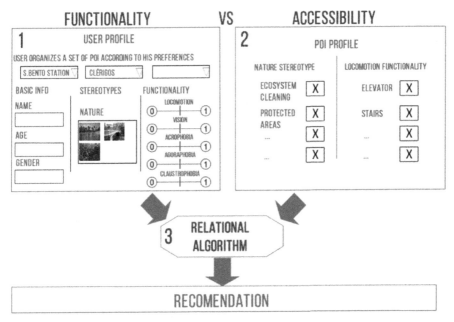

Fig. 1. Describes the approach architecture focusing the work in the dichotomy between user functionality and POI accessibility.

User profiles are created based on a set of information acquired from forms, that is composed by a set of basic information (gender, age and nationality, etc.), stereotype accuracy (level of user considering stereotypes like nature, business or residential tourist defined by Tourism of Portugal), functionality levels in physical and psychological issues (locomotion, vision, acrophobia, agoraphobia and claustrophobia). The importance of each considered topic is defined by a different weight allowing to fine tune the platform.

This proposal assumes a linear modeling technique to design user profiles relation with standardized tourism stereotypes. The implemented platform offers the user the possibility to order a set of images related with different stereotypes (the result order will tell us his preferences). After the ordering task is completed the user classifies the images with a set of tags (awesome, interesting, boring) that in conjunction with the order of each image will allows the extraction of result numerical relation between stereotype and individual user.

Beyond the gathered basic info and stereotype accuracy relation the user answer will their functionality level in each one of the considered physical and psychological issues.

The profile of POI is created using an empirical approach which includes a procedure based on a checklist (conditions for a maximum relation with stereotypes and accessibility level) that evaluates their relation with the above referred tourist stereotype and functionality levels (determining POI capacity to receive tourists with certain levels of physical and psychological issues). This will basically define POI accuracy in a determined scale that is equal to the one used to define user profiles. The final value of each considered item (stereotypes like nature or functionality like locomotion) is obtained with a weighted sum of each characteristics used to define them, using the formula (1).

$$\sum\nolimits_{i=1}^{n} \frac{A*Xi}{n} \tag{1}$$

A –Characteristic weight X –Characteristic value

The relational algorithm referred in the proposed architecture allows the recommendation mechanism to order POI's. This process is divided in 2 steps (stereotype average and functionality average) like it can be seen above.

Relational algorithm to relate stereotype User and POI

```
var STEREO_USER; //Obtained relation between each user and stereotype
var STEREO_POI; //Obtained relation between each POI and stereotype
var  NUM_RELATION_SUM; //numerical relation
var FINAL_VALUE_STEREO; // average value between all stereotypes

FOR i=0 to MAX_STEREOTYPES
     NUM_RELATION= NUM_RELATION_SUM  + (STEREO_USER[i]   *
STEREO_POI[j])
   END FOR
      FINAL_VALUE_STEREO= NUM_RELATION_SUM  / MAX_STEREOTYPES
   END
```

The algorithm that allows the relation between user functionality and POI accessibility has some differences because it is not a simple summation average of the user and POI values like it was shown in stereotype algorithm. Here it is considered that, if in a specific functionality/accessibility issue the user value is higher this will be used in the average calculation instead of the multiplication between user and POI values. The reason for this positive discrimination is that if the user has higher value in a specific physical/psychological issue than the POI it was decided to use the user's value instead of the multiplication between both.

Relational algorithm to relate User functionality and POI accessibility

```
var FUNC_USER; //Obtained relation between each user and
functionality
var ACESS_POI; //Obtained relation between each POI and accessibility
var  NUM_RELATION_SUM; //numerical relation
var FINAL_VALUE_FUNC_ACCESS; // average value between all stereotypes
var COUNTER: 0;

FOR i=0 to MAX_FUNCIONALITY_LEVELS
   if FUNC_USER[i] > ACCESS_POI[i]
     NUM_RELATION= NUM_RELATION_SUM  + ACCESS_USER[i]
        COUNTER= COUNTER +1
   else
        NUM_RELATION= NUM_RELATION_SUM  + (FUNC_USER[i]   *
ACCESS_POI[i])
   END FOR
      FINAL_VALUE_FUNC_ACCESS = NUM_RELATION_SUM  / MAX_STEREOTYPES
   END
```

The recommendation results produced by the system are compared with a pre-organized list of POI made by each user when it firstly enters the system (it is assumed that the user will respect their own characteristics). This will allow checking and concluding if the generated results are accurate and if the new data added is important to this kind of systems.

The developed tools will be evaluated by users feedback obtained in a questionnaire.

4 Conclusions and Future Work

The main goal is to prove that functionality levels should be considered in user profile creation because they can be of the utmost importance in defining what a user can or cannot do when is visiting a specific POI. With this approach it is aimed to prove that this can be the next step in user profile creation regarding new information related with user's physical and psychological functionality levels.

In the next steps the recommendation results will be compared and evaluated to verify which differences are detected between the pre-organized list and the recommendation results. Furthermore some good practices for accessibility standards defined by Tourism of Portugal [32] are being analyzed, in order to evaluate our approach effectiveness.

Acknowledgments. POCI, POSC and COMPETE for their support to GECAD unit.

References

1. Kobsa, A. Generic User Modeling Systems. User Modeling and User-Adapted Interaction. Kluwer Academic Publishers, 49-63 (2001)
2. Kobsa, A. User Modeling: Recent Work, Prospects and Hazards (1994)
3. Porter, J., Watch and Learn: How Recommendation Systems are Redefining the Web (2006)
4. Berka, T., Plößnig, M. Designing Recommender Systems for Tourism, in ENTER 2004.
5. Schafer, J. Ben, Konstan, Joseph and Riedl, John. 1999. Recommender Systems in E-Commerce (1999)
6. Felfernig, A., et al. 2007. A Short Survey of Recommendation Technologies in Travel and Tourism. In OGAI Journal (Oesterreichische Gesellschaft fuer Artificial Intelligence) 25(7). 17-22(2007)
7. Katerina Kabassi. 2010. Review: Personalizing recommendations for tourists. Telemat. Inf. 27, 1 51-66 (2010)
8. Tedlow, Richard S. 2000. Exploring Marketing with Delta Airlines as a Case Study (2000)
9. Faria A., Almeida A., Martins C., Gonçalves R., and Figueiredo L. Personality traits, Learning Preferences and Emotions. In *Proceedings of the Eighth International C* Conference on Computer Science & Software Engineering* (C3S2E '15). ACM, New York, NY, USA, 63-69.(2015)
10. Rich, E. 1979. User Modeling via Stereotypes (1979)
11. Zukerman, I., Albrecht, D. Predictive Statistical Models for User Modeling. In User Modeling and User-Adapted Interaction. Volume 11. 5-18 (2001)

12. Hinze, A.,Buchanan, G. "Cooperating services in a mobile tourist information system." Meersman, R., et al. (eds), Proc. On the Move to Meaningful Internet Systems 2005: OTM 2005 Workshops, LNCS 3762, 12-13. Springer-Verlag, Berlin (2005)
13. Ambite J, Knoblock C., Muslea, M., Minton, S. Heracles: Hierarchical Dynamic Constraint Networks for Interactive Planning (2003)
14. Parle, E., Quigley, A. Proximo, Location-Aware Collaborative Recommender (2006)
15. Espinoza, F., Persson, P., Sandin, A., Nyström, H., Cacciatore. E. & Bylund, M. GeoNotes: Social and Navigational Aspects of Location-Based Information Systems, in Abowd, Brumitt & Shafer (eds.) Ubicomp 2001: Ubiquitous Computing, International Conference Atlanta, Georgia, September 30 – October 2, Berlin: Springer, 2-17 (2001)
16. Biuk-Aghai. R. 2003. MacauMap: Tourism-Oriented Mobile GIS Application. Proceedings of Map Asia 2003, Kuala Lumpur, Malaysia: GIS Development, 1-8 (2003)
17. Höpken W., Fuchs M., Zanker M., Beer T., Eybl A., Flores S., Gordea S., Jessenitschnig M., Kerner T., Linke D., Rasinger J., and Schnabl M. etPlanner: An IT Framework for Comprehensive and Integrative Travel Guidance. In Proceedings of ENTER. (2006)
18. Krosche J., Baldzer J., Boll. S. MobiDENK-Mobile Multimedia in Monument Conservation, In IEEE Multimedia, vol. 11, no. 2, 72-77 (2004)
19. Klante P., Krosche J., and Boll S. AccesSights - A Multimodal Location-Aware Mobile Tourist Information System. In Proceedings of the ICCHP, Paris, France (2004)
20. Simcock T., Hillenbrand S.P. and Thomas B. Developing a location based tourist guide application. In Proceedings of the Australasian information security workshop conference on ACSW frontiers 2003, Chris Johnson, Paul Montague, and Chris Steketee (Eds.), Vol.
21. Australian C. Society, Inc., Darlinghurst, Australia, 177-183 (2003)
21. Schneider, J., Schröder, F. The m-ToGuide Project-Development and Deployment of an European Mobile Tourism Guide. Heidelberg, Germany (2004)
22. Cheverst K., Mitchell, K. and Davies Nigel. The role of adaptive hypermedia in a context-aware tourist GUIDE. Commun. ACM 45, 47-51 (2002)
23. Poslad S, Laamanen H., Malaka R. Nick, A., Buckle P., Zipf. A. CRUMPET: Creation of User-friendly Mobile Services Personalised for Tourism. In Proceedings of the 2nd International Conference on 3G Mobile Communication Technologies (2001)
24. Abowd, D., Atkeson, G., Hong, J., Long, S., Kooper, R., Pinkerton, M. Cyberguide: A mobile context-aware tour guide. Wireless Networks. 421–433 (1997)
25. Pashtan, A., Blattler, R., Heusser, A., Scheuermann, P. CATIS: A context-aware tourist information system, in: Proceedings of IMC 2003, Rostock, Germany (2003)
26. Malaka, R., Zipf. A. DEEP MAP ñ Challenging IT Research in the Framework of a Tourist Information System. In: D. Fesenmaier, S. Klein and D. Buhalis, eds.: Information and Communication Technologies in Tourism 2000. Wien, New York, Springer, 15-27 (2000)
27. Coelho, B., Figueiredo, A., and Martins, C. Tours Planning Decision Support. In ISCIES09. Porto. Portugal. (2009)
28. Lucas Joel P., Luz Nuno, Moreno María N. Anacleto Ricardo , Figueiredo Ana Almeida, Martins Constantino, "A hybrid recommendation approach for a tourism system", Expert Systems with Applications 40 3532–3550 (2013).
29. Wan-Shiou Yang and San-Yih Hwang. iTravel: A recommender system in mobile peer-to-peer environment. J. Syst. Softw. 86, 12-20 (2013)
30. Noguera J. M., Barranco M. J., Segura R. J., Martínez, L A mobile 3D-GIS hybrid recommender system for tourism. Inf. Sci. 215), 37-52 (2012)
31. Colomo-Palacios R., García-Peñalvo F., Stantchev V., Misra S, Towards a social and context-aware mobile recommendation system for tourism, Pervasive and Mobile Computing (2016)
32. Tourism of Portugal, http://www. turismodeportugal.pt

Identification of visually impaired users for customizing web pages on the Internet

Juan Peraza[1] , Yadira Quiñonez[1], Carmen Lizarraga[1], José Ortega[2], Monica Olivarría[1],

[1] Universidad Autónoma de Sinaloa, Facultad de Informática Mazatlán, Av. de los deportes S/N, Mazatlán, Sinaloa, México.
{jfperaza, yadiraqui, carmen.lizarraga, m.olivarria}@uas.edu.mx
[2] Universidad de Granada, Campus Universitario Cartuja S/N, Granada, España.
{jaorte}@ugr.es

Abstract. There are some people with disabilities who try to surf on the Internet every day using tools that help them better visualize the web pages as augmentative systems and listen the web pages containing text through audio descriptors. For this population, this study tries to propose a change in the guidelines of the W3C Web Accessibility for sites that wish to provide access to this sector of the population. This research focuses on the disability of blindness, the proposal is based on the identification of blind people through an update that would apply to an existing variable that is issued by the client's browser, in that way the web servers can read its content and display an accessible web page.

Keywords: Internet, WWW, Visual disability, W3C.

1 Introduction

With the creation of the Internet since it emerged the ARPANET in the late 60's Leiner B. et al [1] has grown exponentially becoming the way most commonly used to transfer data such as voice, wire transfers, various types information, etc. This transfer of information is carried out by people through any device connected to the Internet, of which 20% have some form of disability as auditory, visual or mental problems, WebAIM [2]. According to the World Health Organization there are 285 million visually impaired people of whom 39 million are blind. Of this amount, 82% are 50 years or older, World Health Organization [3]. This means there is a 18% equivalent to more than 7 million people, that are under 50 years of age, these can join to a productive or academic life. But there is a problem, these people have great difficulty in accessing information stored on sites like electronic magazines, online newspapers or any website that stores information. They can't read the information stored on web pages, they need to be shown a special page that can describe audio information.

This research proposes an update based on the necessity of improve the access to information to people with blindness. This involves several changes by the W3C, developers of web browsers and design of websites on the Internet, so that the inclusion of greater number of people is achieved to information offered by nearly 700 millions of websites on the Internet according to Internet statistics Live Stats [4].

1.1 Current state of accessible pages

There are guidelines for creating accessible sites offered by the W3C (World Wide Web Consortium), these are guidelines that serve responsible for designing websites to designing attached to the standards of the W3C sites. These serve as a basis for a site it can be evaluated by the WAI (Web Accessibility Initiative) and be considered as a site accessible. WAI Site [5].

The W3C is responsible for developing standards governing the orderly growth of the Web and through this the necessary protocols exist for a user, through their browsers to access content stored on Web pages. Taking into consideration that visiting a website is usually transparent and easy for the user, behind that simple step there many years of research and development by the W3C (World Wide Web Consortium), from the HTML language that is most used communication protocols between the web server and browser; among others protocols and processes.

Addressing the issue of web accessibility W3C establishes the "WAI" (Web Accessibility Initiative or Web Accessibility Initiative) whose purpose is to facilitate access to web content people with any type of disability through standards and guidelines for the proper design accessible websites, Shawn L. [6].

The issue of web accessibility has been around since at least 15 years and among the projects W3C when he published the opening of the POI (International Program Office for Web Accessibility Initiative) in 1997 with the purpose of promoting and achieving Web functionality for people with disabilities. During that presentation Tim Berners Lee said.

"The power of the Web is in its universality. Access by everyone regardless of disability is an essential aspect", Berners L. [7].

One goal of this research is to propose an improvement to the guide that regulates the W3C, to design accessible websites as this guide is a great help for web masters and even more for people with disabilities but focuses only on the visual appearance of websites.

At present there are few tools that function as audio description texts containing academic information portals, an example is "ReadSpeaker". This works 100% online, is a good reference but is commercial and costs, the second example is "vozMe", you must install a plugin in the browser to use, it detracts portability. ReadSpeaker [8], Cano D. [9].

The operation of most of the previously analyzed tools are mainly based on the download and installation prior to operate the audio description, being almost all commercial software requires the purchase of a license. Among the objectives of this research it is that the end user does not have the worry to download and install software for audio to describe the text. This process should be transparent to the end user where in addition to the impact it will have on user satisfaction, facilitate access to content stored on a website.

This work has a social purpose to provide blind people universal access to great content of information that exists today stored in academic sites, audio describing the text and images that are contained through sites designed specifically for blind people who may be detected when this person enters the website.

1.2 Accessing web pages

To explain the proposal of this work is necessary to know the process to access a webpage, this is well known by everyone from the point of view of the end user. This must have a device with internet connection (can be a desktop or laptop, tablet or cell phone) you need a web browser and enter the URL in the address bar to access the site to view. Basic Computer Tutorial, UNAM [10].

When you access a website, what actually happens is that the web browser helps us to connect to a web server through port 80 using the HTTP protocol. This web server serves us our request and displays the website depending on the domain name you're asking.

It may be the case that a web server to host many websites, this means that the virtual server can be configured domains or virtual hosting websites and display visually different.

The term "Virtual Hosting" refers to operate more than one web site (such as www.company1.com and www.company2.com) on a single machine. Virtual websites may be "based on IP addresses", which means that each site has a different IP address, or "based on different names", which means that with a single IP address are running websites with different domain names. The fact that they are running on the same physical server is not apparent to the user visiting these websites," Apache HTTP Server Documentation [11].

1.3 Identification of visually impaired users

The identification user process could be very simple, but we can break down this process into several parts. Starting with the process Web client, it emits a variable called HTTP_USER_AGENT which is defined according to the official manual programming language "PHP":

"Contents of the User-Agent: header from the current request, if any. It consists of a string that indicates the user agent used to access the page. A typical example is: Mozilla / 4.5 [en] (X11; U; Linux 2.2.9 i586) ", PHP Official Manual [12].

Another definition for the User-Agent described on RFC 1945 is:
"The client which initiates a request. These are often browsers, editors, spiders (web-traversing robots), or other end user tools. The User-Agent request-header field contains information about the user agent originating the request. This is for statistical purposes, the tracing of protocol violations, and automated recognition of user agents for the sake of tailoring responses to avoid particular user agent limitations. Although it is not required, user agents should include this field with requests. The field can contain multiple product tokens and comments identifying the agent and any subproducts which form a significant part of the user agent. By convention, the product tokens are listed in order of their significance for identifying the application.

User-Agent   = "User-Agent" ":" 1*(product | comment)

Example:

User-Agent: CERN-LineMode/2.15 libwww/2.17b3" Berners-Lee T.(1996) et al. [13]

This variable is issued by the client, if we analyze it, we find that the values that make up this chain are separated by ";". These values are read from the server side, where web pages displayed to users are stored. Some websites currently analyzing this variable and print a personalized website depending from what type of device you are connecting the customer.

It is important to emphasize that today already can display a personalized and stylized website for each type of device the user is using to access the website.

How is this possible? this is possible through parsing the User-Agent String using the Product Tokens.

"These are used to allow communicating applications to identify themselves by software name and version. Most fields using product tokens also allow sub-products which form a significant part of the application to be listed, separated by whitespace. By convention, the products are listed in order of their significance for identifying the application.

    product   = token ["/" product-version]
    product-version = token

Examples:

    User-Agent: CERN-LineMode/2.15 libwww/2.17b3
    Server: Apache/0.8.4" Berners-Lee T.(1996) et al. [13]

These pages are displayed depending on the device used to display, these pages are called progressive, which have certain characteristics that must meet to be considered a "Progressive Web App". These characteristics according to google developer guide are: [14]

– Progressive - Work for every user, regardless of browser choice because they're built with progressive enhancement as a core tenet.
– Responsive - Fit any form factor: desktop, mobile, tablet, or whatever is next.
– Connectivity independent - Enhanced with service workers to work offline or on low quality networks.
– App-like - Feel like an app to the user with app-style interactions and navigation because they're built on the app shell model.
– Fresh - Always up-to-date thanks to the service worker update process.
– Safe - Served via HTTPS to prevent snooping and ensure content hasn't been tampered with.
– Discoverable - Are identifiable as "applications" thanks to W3C manifests and service worker registration scope allowing search engines to find them.

- Re-engageable - Make re-engagement easy through features like push notifications.
- Installable - Allow users to "keep" apps they find most useful on their home screen without the hassle of an app store.
- Linkable - Easily share via URL and not require complex installation. [14]

Another definition of Progressive Web App (PWA), It stands for Progressive Web App, and it lets developers build a mobile website can perform That super-fast and behave just like an app. "They are a better way to enable a website to work more like a native app installed," Said Aaron Gustafson [15].

Another term that we must consider is Responsive Pages which can adapt the visual appearance using HTML or CSS to fit the screen resolution of any device used to access the website. This technology provides the user get the best display of a website on any mobile device, whether a cell phone, tablet or desktop PC. Responsive Web Design Introduction [16].

K. Knight [17], defines this concept as suggests that approach is the design and development that should respond to the user's behavior and environment based on screen size, platform and orientation. The practice consists of a mix of Flexible grids and layouts, images and an intelligent use of CSS media queries.

We'll take a question, what if we could also display a custom website and stylized for certain types of users suffering from a visual impairment?, This idea can be realized only have to find a way to know if a client suffers from a visual impairment. One way to carry out such identification is explained below.

2 Methodology

Once we know the functionality of HTTP_USER_AGENT, what is intended to propose this work is to update the web browsers, modifying the variable HTTP_USER_AGENT, adding a configuration parameter called "speab" referring to "special ability". The name used for this variable is only a prototype name to refer to a new variable.

As was mentioned, this research is a proposal to improve the guidelines of the W3C for the benefit of people who suffer some weakness, this proposal initially focuses on people with blindness.

Within the design of this proposal, the variable "speab" can be changed during the setup process to install the web browser or to perform initial configuration of a computer, regardless of operative system use. This variable would become a parameter to be added to the variable "HTTP_USER_AGENT".

Through this variable (HTTP_USER_AGENT), from the webserver side you can detect if a user connects from a mobile device or from a desktop computer. Boxall J. [18].

With this parameter, you can see if the client suffers from a problem, in this paper, we will focus on detecting people with visual disability, proposing that the parameter "speab" will add some characters that refers to visual disability leaving "speab-v" (special visual capacity), if the person has visual and auditory weakness would add the letter "a" being "speab-va". An example of how the variable HTTP_USER_AGENT would be: "Mozilla / 4.5 [en] (X11; U; Linux 2.2.9 i586;

speab-va)" with the last parameter of the chain, we are specifying that the user suffers from a special visual ability.

It is important to mention the character to separate is "-" horizontal character line used as a token to separate the parameters, in this way the parser can read the parameters that can be declared in the future.

In this way the website can parse the string containing the variable and detect whether the user is visiting the site has a disability. The website, knowing this information can print a page with a special design and appropriate tools depending on the disability which the user has.

2.1 Parsing USER_AGENT Variable

There are some ways to parse the variable to detect the new parameters that we added. One of these is using PHP code to split the variable by tokens for example:

Using this USER_AGENT as a example: Mozilla / 4.5 [en] (X11; U; Linux 2.2.9 i586; speab-v

Table 1. Splited by tokens.

Token Number	Content
1	Mozilla
2	4.5
3	[en]
4	X11
5	U
6	Linux 2.2.9 i586
7	speab-v

In this example on Table 1, we can see that there is a token number "7" with "speab" variable declared, with "v" assigned. This means that the user who has this USER_AGENT, declares that has a visual impairment and need a special designed web page.

This parsing process could be a function on header section of the principal index of a webpage what could be checking every user accessing the webpage and if is needed the webserver redirects the user to a special webpage designed depending on the special condition.

This leads to a web development where you can develop various applications that can assist users with visual disabilities, imagine a blind person making a purchase of a service through an accessible web page that displays tools talk around with an operator remotely assisting him.

A survey was applied to 20 blind people were asked if they would like to visit a Web page describing the audio content, 100% of respondents answered "yes". Also they were asked how often they dismiss a website for failing to listen to the description of the content with the tool they have installed on your pc. of the following possible answers "a) never b) rarely c) very frequent", 100% answered "C" (very frequent).

3 Conclusions

This work shows the need to improve certain guidelines in W3C. A proposal is exposed based on the needs of people with disabilities, in order to detect such users and try to make websites more accessible for them.

This is something that the CONAPRED (national council to prevent discrimination) and CONADIS (National Council for the development and inclusion of people with disabilities) have been working in Mexico but there is still work to be done. Even the accessibility has been proposed in federal law in Article 32 in paragraph no. 2 of the General Law for the Inclusion of Persons with Disabilities, which says:

"II. Promote the use of Mexican Sign Language, Braille, and other modes, means and formats of communication and access to new systems and information technology and communications, including the Internet ". General Congress [19].

The Internet offers many opportunities for people with disabilities that are not available through any other resource, offers independence and freedom. However, if a website is not created based on the standards of web accessibility, it excludes a segment of the population that can benefit a lot from the Internet.

This is because many of the activities they perform daily the could carry out through their computer. Most web masters are not intended to exclude persons with disabilities, but until today, there are very few who develop a site accessible under the guidelines of the W3C. Organizations and designers to raise awareness and implement accessibility, will ensure that your content will be available to a wider population.

With this research we are expanding the spectrum of how websites in the future perhaps not too distant be composed, do not consist only of the visuals but also of the auditory part and why not (as mentioned in section above) to people who are behind a desk waiting for visually impaired customers to remotely assist with a critical process as it can be a purchase through an accessible website.

References

1. Leiner B., Cerf V., Clacrk D., Kahn R., Kleinrock L., Lynch D., Postel J., Roberts L., Wolff S. (2012). Brief History of the Internet. Internet Society, http://www.internetsociety.org/internet/what-internet/history-internet/brief-history-internet [Visited 17-07-2016].
2. WebAIM, (2014). Introduction to Web Accessibility, http://webaim.org/intro/ [Visited 18-07-2016].
3. World Health Organization, (2014), Visual impairment and blindness, http://www.who.int/mediacentre/factsheets/fs282/en/, [Visited 28-08-2016].
4. Internet Live Stats, (2014). Real Time Statistics Project, http://www.internetlivestats.com/ [Visited 17-07-2016].
5. WAI Site (Web Accessibility Initiative), https://www.w3.org/WAI/ [Visited 17-07-2016].
6. Shawn L. (2005), "Web Accessibility Initiative", Ed. W3C, http://www.w3c.es/Traducciones/es/WAI/intro/accessibility [Visited 19-07-2016].

7. Berners L. (1997), "World Wide Web Consortium Launches International Program Office for Web Accessibility Initiative", http://www.w3.org/Press/IPO-announce [Visited 19-07-2016].
8. ReadSpeak Official Documentation, http://www.readspeaker.com/readspeaker-docreader/, [Visited 03-09-2016].
9. Cano D., "Nota Legal" http://www. vozme.com/, [Visited 03-09-2016].
10. Tutorial Cómputo Básico UNAM (2006). Como entrar a una Página. Universidad Nacional Autónoma de México, http://borges.dgsca.unam.mx:8080/tutorialcomputo/ index.jsp?pagina=correo&action=vrArticulo&aid=64 [Visited 18-07-2016].
11. Apache HTTP Server Documentation, http://httpd.apache.org/docs/2.0/es/vhosts/ [Visited 03-09-2016].
12. Manual Oficial PHP (2014). The PHP Group Ed., http://php.net/manual/es/reserved.variables.server.php [Visited 18-07-2016].
13. Berners-Lee T., Fielding R., Frystyk H., "Request For Comments: 1945", https://tools.ietf.org/html/rfc1945#section-10.15, May 1996.
14. Google Developers, https://developers.google.com/web/updates/2015/12/getting-started-pwa [Visited 18-07-2016].
15. Aaron Gustafson (2016), The Web Should Just Work for Everyone, https://aaron-gustafson.com/notebook/the-web-should-just-work-for-everyone/ [Visited 18-07-2016].
16. Responsive Web Design Introduction, http://www.w3schools.com/css/css_rwd_intro.asp [Visited 18-07-2016].
17. K. Knight (2011), Responsive Web Design: What It Is and How To Use It, Smashing Magazine, https://www.smashingmagazine.com/2011/01/guidelines-for-responsive-web-design/ [Visited 22-07-2016].
18. Boxall J. (2009). Mobile Device Detection, http://notnotmobile.appspot.com/ [Visited 18-07-2016].
19. Congreso General (2011), "Artículo 32 de la Ley General Para La Inclusión De Las Personas Con Discapacidad", 2011.

Author Index

Printed in the United States
By Bookmasters